BENJAMIN HENRY LATROBE & MONCURE ROBINSON:

The Engineer as Agent of Technological Transfer

REGIONAL CONFERENCE IN ECONOMIC HISTORY MAY 17, 1974

Edited by Barbara E. Benson

Cover   "Sketch for a design of an Engine house and Wateroffice
in the City of Philadelphia," by Benjamin Henry Latrobe,
1799 (The Papers of Benjamin Henry Latrobe, Baltimore).

TABLE OF CONTENTS

ILLUSTRATIONS

INTRODUCTION

Brooke Hindle

The transfer of technology has recently emerged as an important field of historical inquiry. It is a key factor in creativity and development in technology. However, it is currently perceived as being important because it is highly relevant to efforts directed toward improving the lot of underdeveloped nations. Recent research demonstrates that both historians and policy planners have much to gain as the inquiry advances.

The transfer of technology, by definition, is an international phenomenon. Transfer within a region is usually thought of as the diffusion of technology, a related but somewhat different process. In 1968, ICOHTEC, the International Cooperation in History of Technology Committee, was established to increase the attention given the history of technology. It immediately identified the transfer of technology as the most important single area to which it should devote its primary efforts. The first result was a separate conference on "L'Acquisition des Techniques par les Pays Non-Initiateurs." This conference, held at Pont-à-Mousson in 1970, was followed by a second on the same topic, but covering a later time period, held in Yugoslavia in 1972. At the International Congress of the History of Science held in Japan in 1974, the history of technology bulked large, and two of the eight symposia were concerned with the transfer of technology. This time the focus was upon the transition from non-initiator countries to initiator countries.

Within this country the Society for the History of Technology and other historical societies have demonstrated some interest in the transfer of technology, but thus far activities lack focus and direction. Conference sessions on the subject suffer from an approach combining unrelated viewpoints and unrelated areas. So far little has been published beyond an occasional article or tangentially interested books.

For this reason, the conference held at Eleutherian Mills Historical Library, from which this book has proceeded, has a special importance. It offers a detailed examination of one method of technology transfer covering a limited period of time, the early part of the nineteenth century, and a limited geographical region, primarily the central seaboard states. Moreover, the papers represent an important identity of viewpoint. They employ a biographical approach to one class of carriers of technology--the engineer. The human carrier has always been one of the primary agents of transfer, and in early nineteenth century American history, he was the dominant agent.

While the biographical approach is fundamental in technology, examples of its misuse litter the discipline's historiography. This is reflected in the presentation of technology as eponymic character and the belief that invention and technology are synonymous. In the popular journalistic view, the great hero inventors placed the milestones: Eli Whitney invented the cotton gin and interchangeable parts; Elias Howe invented the sewing machine; Samuel Morse invented the telegraph; Cyrus McCormick invented the reaper; and Samuel Colt invented the revolving pistol. In such accounts there is little or no consideration of the transfer of technology. This conference and this book use biography correctly and transcend the eponymic view of technology.

Much more important than eponymic distortion, in the case of Benjamin Henry Latrobe, has been the attention to only one of his many areas of achievement. After gaining both his general and technical education in Europe, Latrobe came to this country and distinguished himself in several fields of endeavor. Most scholars, including his finest biographer, however, have presented him only as an architect and have evaluated him in that single area. This is merely part of the story, for Latrobe made important contributions to construction, steam engine development, water supply systems, and even mining

and metal processing. An examination of all the aspects
of his career reveals a wholly different profile. Moncure
Robinson has been even less fortunate. Unlike Latrobe's
architectural efforts, Robinson's engineering feats were
not interesting to publishing scholars of any description.

There are, of course, many mechanisms by which
technology is transferred. These include the importation
of the actual machines; the transmission of blueprints,
specifications, or patent descriptions; and the transmission of books and treatises. Because high science technology today demands a competent science community to
receive and perpetuate it, it has sometimes been assumed
that the most important ingredient in transferring technology was the transfer of science and scientists. At least
in antebellum America, there was little validity to this.
In that period, the primary agent was the individual. The
two primary human carriers are represented here: the
European who learned the technology abroad and brought
knowledge of it with him to this country, and the American
who went abroad to learn the technology.

The character of the barriers against the transfer
of technology which existed when Latrobe came to America
and when Robinson went to Europe is significant. Great
Britain was the center of the classical Industrial Revolution, and she perceived that this complex of technological
and economic changes gave her enormous advantages over
the rest of the world. She therefore enacted laws to pro-
hibit the transfer of the associated technologies beyond the
British Isles. These prohibitions were based upon the as-
sumption that there were two primary agents of transfer.
The first was the machine itself, which lay at the center
of the Industrial Revolution. Its exportation, therefore,
was prohibited, except by special license. The second
mechanism was the human carrier; to control that, the
emigration of artisans was prohibited.

The barriers against the transfer of technology
proved ineffective, not because they were raised against
the wrong agents of transfer, but because they were cir-

cumvented. British artisans easily escaped the restric-
tions upon emigration and went, especially to the United
States, with very little hindrance. As for machines, the
Americans quickly passed the need for importing British
machines, so that by the time that ban was raised, in
1843, exports to this country were inconsequential.

The term "engineer" was not used in the modern
sense when the British barriers were erected, but it did
emerge during Latrobe's and Robinson's lifetime. Engineers constituted a different category of human carrier
from artisans and mechanics. The engineer was a visible
agent, while the skilled mechanic remained largely anony-
mous and even today cannot be traced easily. Yet, the
effective execution of the engineer's projects depended
heavily upon skilled mechanics capable of building the
various elements constituting a technological system,
such as a railroad. The invisible migration of artisans
underlay the efforts of the engineers. At its best, the
skill of the artisan was a creative resource which could
be applied to problem solving. It looked toward adaptation and further development--not merely to the perpetua-
tion of the technology being transferred.

The engineer transferred two important capabilities.
First, he brought specific knowledge of one or more technologies. He knew the processes; he knew the manner of
planning and building; he had experience with successful,
developed technological systems. This is what is usually
thought of in the transfer of technology, but it is not the
most important capability. The transfer of railroad technology to America did not consist in building here exactly
the same sort of rails and roadbeds and running English
locomotives on them. Rather it was a question of transferring an almost living organism with a capability of
adaptation, development, and growth. This creative
ability to adopt a new technology and to develop it in a
new region was the most important thing carried by the
engineer.

What the engineer transferred was intangible and
elusive. Edward C. Carter II writes of it when he says

that the "technological freight is only as good as the man who transports it. " Carter believes that Latrobe brought with him more than specific information: he brought an interest in a variety of technologies and applications of technology; he brought skill, industry, outstanding ability, and inventiveness. In thinking about the same question, Darwin Stapleton notes that the engineer transferred attitudes and standards in addition to skills. He brought an acquaintance with books related to technology and with other individuals who could constitute a part of the environment of growth.

The transfer of technology very clearly requires the establishment of a technological community. Therefore, the role of engineers as agents of transfer consists not merely of building and adapting, but of instructing. Latrobe and Robinson participated in the diffusion of knowledge within this country, and Latrobe was especially distinguished for his outstanding students. Both men were among the charter members of new technological communities.

The case studies by Carter and Stapleton and the fine analysis and extension of John B. Rae are remarkably successful in addressing the same questions. Together they illuminate our understanding of one class of the agents of technological transfer. This new light is directed toward the center of the process. The book advances our understanding another step and demonstrates its importance by the stimulus it provides.

Benjamin Henry Latrobe, oil painting by Rembrandt
Peale, <u>ca</u>. 1816 (collection of Mrs. Gamble Latrobe).

THE ENGINEER AS AGENT OF TECHNOLOGICAL TRANSFER: THE AMERICAN CAREER OF BENJAMIN HENRY LATROBE, 1796-1820

Edward C. Carter II

From the moment of his arrival in Virginia in March 1796 until his death a quarter of a century later in New Orleans, Benjamin Henry Latrobe (1764-1820), who often signed his plans "Engineer," was deeply involved in the study and advancement of American technology. He was greatly interested in the manner in which wharves were built in Norfolk, and he filled several pages of his first Virginia journal with sketches and notes concerning "the invention of Mr. Owen, a Welshman." Our last glimpse of the great architect and engineer shows him returning, exhausted but unafraid, to the construction of the New Orleans waterworks suction pipe as the level of the Mississippi gradually fell and yellow fever spread throughout the city. When death claimed Latrobe on September 3, 1820, he was on the verge of completing the waterworks, a project he had entered into a decade previously with his talented son Henry Sellon Boneval Latrobe (1791-1817). Ironically, Latrobe the younger had succumbed to yellow fever just three years earlier.[1] Benjamin Henry Latrobe was the genius of his age and is considered the founder of professional architecture in America. He brought the highest level of European artistic and technical proficiency to this country and was able with hard work and sympathetic insight to adapt those skills to a frontier situation. He built the U. S. Capitol, one of the most important structures in the western world, in a frontier village, Washington.

What was true of his architecture was also true of his engineering. Often both were joined as in the case of the Philadelphia waterworks and the U.S. Capitol. One of the great stories that The Papers of Benjamin Henry Latrobe will reveal is that of the continual transference of European technology to the United States. Hopefully, the publication of our microfilm and letterpress editions will help resolve some of the questions raised by Brooke Hindle

in his perceptive essay "The Exhilaration of Early American Technology."[2]   The purpose of this brief paper is to
outline the many ways in which Benjamin Henry Latrobe,
Engineer, served as an agent of technological transfer
during his American years. It touches on the transfer of
European skills and attitudes to the United States, the
process of adjustment of an advanced technology with a
less developed culture, and the dispersal of technology
within the new nation. The approach followed is frankly
descriptive rather than analytical; however, that description is placed within a framework of the possible range of
technological transfer. Hopefully, it will encourage others
to seek a fuller understanding of Latrobe's role in this
vital aspect of the transit of civilization.

The first problem to be faced is one of definition.
What was an engineer during this period of American history? Did Latrobe conform to this model? While rec-
ognizing that an engineer prior to 1820 might on occasion
engage in other activities, Daniel Calhoun, in his excel-
lent study of the early American engineer, concentrated
on those things which he deemed "typical in his work experience [the engineer's] --on his employment by private
corporations and state governments to direct the building
of transportation projects."[3] Latrobe's career far ex-
ceeded this definition of an engineer as one involved with
the direction of internal improvements. As we shall see,
it adhered to that description of civil engineering put forward in 1828 by the Institution of Civil Engineers in Great
Britain. It described the profession as:

> . . . . that species of knowledge which con-
> stitutes the profession of a civil engineer;
> being the art of directing the great source
> of power in nature for the use and convenience
> of man; of the means of production and of traf-
> fic in states, both for external and internal
> trade, as applied in the construction of roads,
> bridges, aqueducts, canals, river navigation,
> and docks, for internal intercourse and ex-

Philadelphia Public Water Wells

— before 1800

became burdened by population

New Waterworks (Latrobe: 1798/99
promoted prior to construction
with engraved pamphlet

Phillyh2o.org

Phillyh2o.org

> change: and in the construction of ports, har-
> bors, moles, breakwaters, and light-houses;
> and in the art of navigation by artificial power,
> for the purposes of commerce; and in the con-
> struction and adoption of machinery; and the
> drainage of cities and towns. [4]

Latrobe was also, because of his extensive involvement with steam power, what would later be termed a mechanical engineer.

In the two decades between his first great engineering project, the Philadelphia waterworks, and his final effort in New Orleans, Benjamin Henry Latrobe must have had the largest and most varied engineering practice, both active and consultative, of any man in America. Calhoun refers to his Da Vinci versatility. [5] The geographical distribution of his work was impressive: New York, eastern and western Pennsylvania, Delaware, Maryland, Virginia, Ohio, and Louisiana. I suspect that the project dollars expended under his direction (including the structural, acoustical, and heating portions of the U. S. Capitol work) far surpassed that of any single engineer before the age of railway construction.

As a general American historian, it appears to me that there are at least five ways in which an engineer may serve as agent of technological transfer. It is within this range of possibilities that I intend to discuss Benjamin Henry Latrobe's engineering activities. First, there is the actual transfer of the engineer's unique skills and knowledge, the mysteries of his calling, from one physical location to another, for our purposes from a sophisticated technology to a less advanced one. [6] Second, of great importance is the engineer's ability to adapt his skills and techniques to a new environment. Technology is worthless if it cannot be made functional. Adaption may require simplification of transferred skills and also involve borrowings from native technology. Third, the engineer, on occasion, serves as a creator of technological momentum. [7] Because of the scope and nature of his work, new projects are activated. A favorable official and public climate of opinion emerges which supports large scale integrated technological planning. Fourth, there is the engineer who because of his high competence and willingness and ability to communicate becomes a center of technological dissemination. Fifth, the engineer may be the agent of continuing technological transfer by serving as a conduit for the flow of ideas and techniques from one society to another.

What was the nature of Latrobe's European education and professional experience; how were these skills translated to the American scene? Benjamin Henry Latrobe received the finest education available in the Anglo-American world. He passed through the Moravian school system in England and Germany, fortunately avoiding Oxford and Cambridge. His father, Benjamin Latrobe (1726-1796), was a leading Moravian minister, Headmaster of Fulneck School, a superb administrator, a scholar and accomplished musician, and a friend of both Dr. Charles Burney and Dr. Samuel Johnson. The younger Latrobe acquired not only a sound classical education but also mastered modern languages (German, French, Italian, Spanish, and contemporary Greek), modern history, theology, mathematics to calculus, physical science, and natural science (biology and geology). He lived in Europe from 1776 to 1784, mostly in Germany; at the end of this period, he toured France and Italy. By then, he had decided to become an architect. Latrobe returned to London ca. 1784 and for several years studied and worked with the famous engineer John Smeaton (1724-1792). Then ca. 1787-1788 he entered the office of the equally renowned architect S. P. Cockerell (1754-1827) where he rapidly rose to the position of chief draftsman. He worked on the Admiralty Building in Whitehall which was designed by Cockerell. [8]

Latrobe's early engineering career is now just emerging as the collection phase of The Papers of Benjamin Henry Latrobe ends and we evaluate newly discovered records. His first task for Smeaton was to report on the "scouring works" which his preceptor had designed in the East Anglian fen country. [9] From Smeaton, the young engineer learned to produce magnificent architectural and engineering delineations. About this time, he became an accomplished watercolorist. Later between 1791 and 1794, Latrobe worked, probably as divisional engineer, on the Basingstoke Canal. His most important engineering work was his unexecuted commission for the Chelmsford Canal which consisted of "deepening and straightening the Chelmer between Malden and the sea," passing the canal through Malden, and improving that busy town's harbor. [10] We have photographs of his two beautifully executed maps of 1793 and 1794 which are important antecedents of the Susquehanna Survey map of 1802, a major document of early American technological history. In addition, the Essex Record Office at Chelmsford has provided us with Latrobe's estimates for the work and his final report together with recently accessioned documents that detail Latrobe's role in the entire improvement scheme. Charles E. Peterson states that Benjamin Henry Latrobe possessed the full range of technical skills available in the big London architectural and engineering offices when he left England in late 1795. [11] It is now evident that Latrobe had a unique training, varied practical experience, and had planned an important local technological undertaking before he departed for America.

Coupled with these achievements was Latrobe's intense scholarly and theoretical interest in his professions. He had seen the large-scale engineering projects of his own times in Germany, France, and England. His study of the art and archaeology of the past informed him of Egyptian, Roman, and Italian Renaissance building technology. His 1,500 volume library containing many professional books was lost in transit to America, and he spent years attempting to reconstruct it. Distance from European book markets and his personal poverty frustrated this effort. In 1804, Latrobe ordered $400 worth of French engineering and architectural treatises, among them the

works of Jean Nicolas Louis Durand of the École Polytechnique faculty and M. Belidor's L'Architecture Hydraul-
ique.[12] John H. B. Latrobe (1803-1891), the elder son of
Latrobe's second marriage, remembered his father's
small French and German collection of technical works and
the beautiful plates of Stuart and Revett's Antiquities of
Athens. In the year of his death, 1820, Latrobe was seek-
ing basic engineering studies used at the Polytechnique for
his son, then a fourth-year West Pointer.

On occasion, Latrobe's dependence on theory caused
him grief. His plan for the Philadelphia waterworks was
attacked because, among other things, it marshalled support for his theories from the works of "such foreign engineers as Bernoulli, Belidor, and Kaestner . . . ."[13]
On the other hand, his dedication to natural history and
his knowledge of geology was directly beneficial to his canal
work. He coupled a brilliant eye for topography with his
understanding of stratification in making his survey estimates concerning excavation, seepage, building material
resources, and water availability at lock levels. In passing, he noted mineral deposits and their economic potential. In 1818, he declined an appointment as State Engineer of North Carolina. Latrobe's letter to Governor
Joseph Gales analyzes the whole problem of a proposed
canal system in terms of the state's topography and geology.[14]

The range of Benjamin Henry Latrobe's technological activities in America is more than impressive. What
follows is merely a listing of his more important undertakings.[15] Some of these examples of technological transfer were "American firsts;" all were executed with professional skill. Latrobe's engineering commissions began
with internal improvements: the improvement of river
navigation and the construction of canals.[16] Latrobe made
a survey and a report on the navigation of the Appomattox
(1797) and executed the survey and improvement of the
Susquehanna channel from Columbia to Tidewater (1801-
1802). He drew plans for an extension of the Potomac
Canal from Great Falls to the Washington Navy Yard (1802).

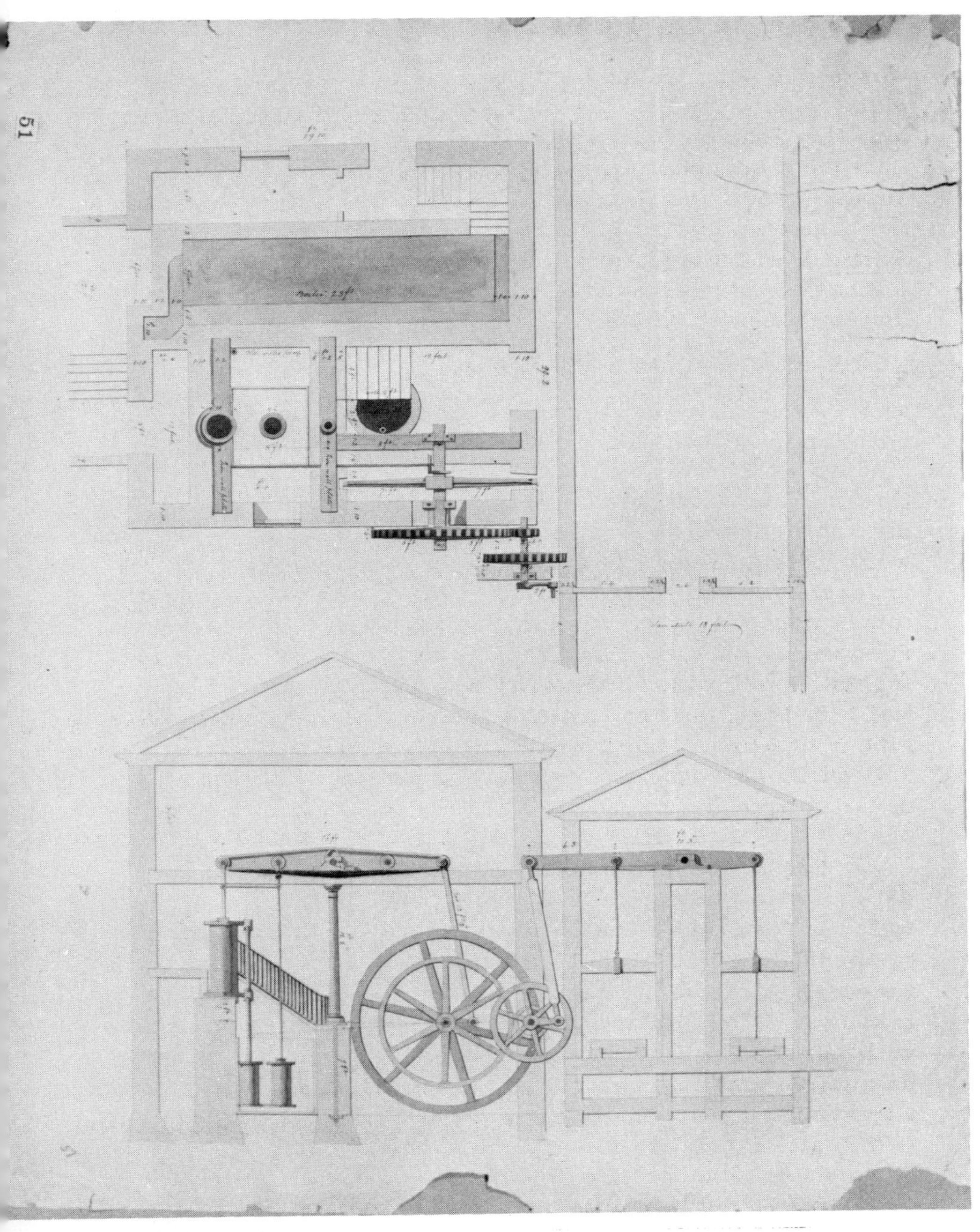

esign of a steam powered sawmill, by Benjamin Henry
atrobe (The Papers of Benjamin Henry Latrobe).

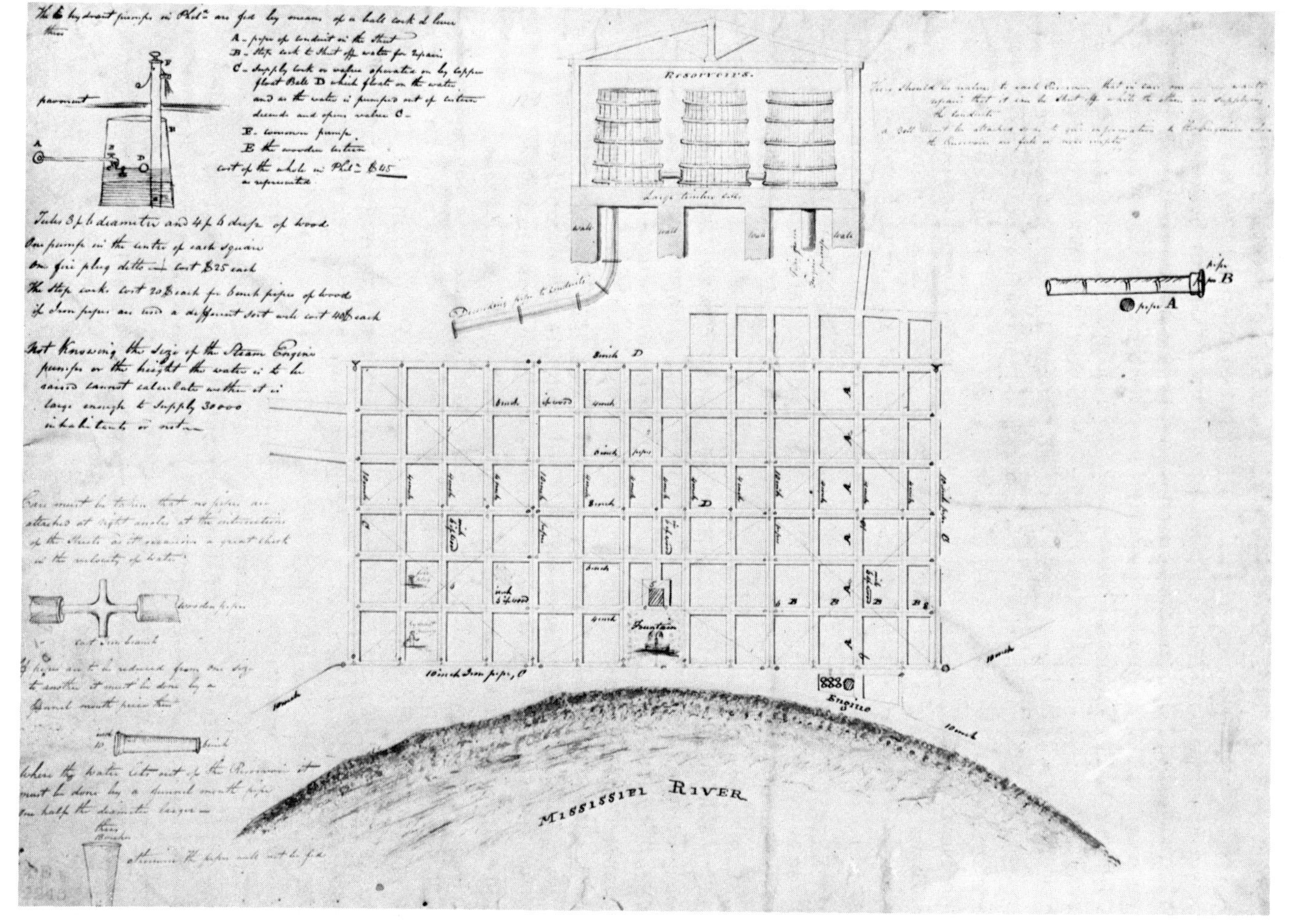

Drawing of reservoirs and street pumps and map of New Orleans waterworks

From 1803 to 1806, Latrobe was the Engineer of the Chesapeake and Delaware Canal. He executed map surveys, wrote
numerous reports, selected a route that William Strickland
(1787-1854) confirmed when the company was revitalized in
the early 1820s, began general excavation, and constructed
a feeder canal to carry materials and supply the main line
with water. Latrobe also planned and built the Washington
City Canal just prior to the War of 1812. For all these projects we have graphic and literary documentation. On three
occasions, he was offered the position of Chief Engineer of
the New York Western Navigation. Latrobe also played a
role in other internal improvement projects as a Commissioner of the National Road and Chairman of the Commissioners of the Baltimore-Washington turnpike. Perhaps
his greatest service lay in his advice to Secretary of the
Treasury Albert Gallatin concerning the creation of a
national internal improvement system of roads, canals,
and navigable rivers. [17]

The second great area of Benjamin Henry Latrobe's
technological work was the utilization of steam power.
This aspect of his career later would be designated mechanical engineering. His correspondence shows he was
well acquainted with the varied modes of employment of
steam power in England. The Philadelphia waterworks
(1799-1801) employed two steam engine pumps; his New
Orleans project one. Latrobe fought with Oliver Evans
(1755-1819) continually and failed to appreciate the potential advantages of the high-pressure engine. He wrote reports on steam engines for the American Philosophical
Society, commissioned and designed steam engines, and,
when Engineer of the Navy, introduced one at the Washington Navy Yard in 1810 that drove a sawmill, a forge
(blower and hammer), and a blockmill. [18] He began construction of two steamboats at Pittsburgh (1813-1814)
after he constructed a unified shipbuilding shop that showed
the same concern for integrated technological planning he
had exhibited in his rebuilding and expansion of the Washington Navy Yard (1804). [19] Carroll W. Pursell, Jr., in

his study of early stationary steam engines, claims Latrobe
was the "leading steam engineer in the country" who, by
coupling "ability and learning and culture, . . . raised the
status of technology and made it a subject of serious consideration." He was the foremost of a band of English engineers whose "frequent immigration kept America abreast
of English developments . . . ."[20]

Benjamin Henry Latrobe's greatest contribution to
American technology was in structural engineering where
he used and developed the "vault as a major element in
American architecture."[21] He used the vault in numerous ways: spanning great spaces, creating beautiful fire-
resistant structures. The Richmond Penitentiary, the Bank
of Pennsylvania, the U.S. Capitol, the Baltimore Cathedral,
and the Baltimore Exchange are all examples of "a brand-
new structural method" that Latrobe introduced into the
United States that not merely spanned great voids but also
was "a major element in architectural design."[22]

The list runs on and on. Benjamin Henry Latrobe
was involved in a wide area of industrial technology, running the "gamut from ironworking to weaving and spinning
machinery."[23] He supplied a steam engine and invented a
new quilling machine for Steubenville textile factories. Continually he pushed the inventions of others (here he was more
typical of a scientist than a technologist). As he had previously done in England, Latrobe experimented with non-
steam power railroads in canal construction.[24]

Latrobe generally succeeded in adapting his technological skills to the American environment. He was open-
minded, liked people, and possessed unlimited optimism,
energy, and physical strength. These qualities helped him
accomplish a prodigious amount of work. Still cultural and
economic conflicts existed between English and native American technology. The basic issue was one of permanence.
As Talbot Hamlin notes, Latrobe favored the English tradition of construction "designed for permanence and low upkeep" while he fully recognized in "the United States low
investments, short amortization periods, and the expecta-
tion of frequent renewal often controlled engineering design."[25]

He tried to bridge the gap between the two attitudes but sometimes found himself frustrated and caused to compromise his
fundamental aims. He was not impressed with the American
custom of employing wood for major structural elements.
Finally, in America, "he was forced to surmount . . . a
general suspicion of theory as theory, a fundamental doubt
of the value of professional advice."[26]

        Technology probably is always altered to some degree
by transference. We do not yet know whether Latrobe sim-
plified his building technology in the United States or whether
he always trained American artisans to meet European standards. He continually refused to allow substandard work on
his projects. While willing to listen to his foreman about
potential laborsaving ideas, his European orientation was
sometimes damaging. Latrobe's rejection of local building practices resulted in the failure of the foundation of the
New Orleans customs house and the Mississippi lighthouse.
In both cases, he imposed unsound theory on conditions with
which he had not yet gained personal experience.

        Latrobe's most serious failures in adapting his tech-
nological expertise to the new world lay in the realm of poli-
tics and human relations rather than that of the actual execution of his projects. He had trouble with boards charged
with the direction of public works from the moment he com-
menced the Richmond Penitentiary. Some jobs went better
than others particularly when he had the protection of a
sympathetic executive such as Thomas Jefferson. The great
Virginian fully comprehended that his Surveyor of Public
Buildings was unrivalled in his profession and took great
pains to educate him in the ways of congressional committees.
True, Latrobe's cost estimates sometimes were low, but
he was executing projects on a scale never witnessed before in America and was continually faced with a scarcity of
quality materials and trained assistants and laborers. The
speed with which he worked was truly amazing, the rebuilding of the U. S. Capitol being a case in point. He was ap-
preciated by the majority of those charged with the construction of the Philadelphia waterworks and the Chesapeake
and Delaware Canal. David Calhoun feels that Latrobe never

grasped the true meaning of the inertia and resistance he
encountered as not being based on republican Americans'
hostility to professional training so much as a widely held
belief that local proprietorship or direct personal economic
involvement in a particular undertaking was the best guaran-
tee of trustworthiness. This "the mere professional could
not offer."[27] In a letter to Volney of 1811, the "Engineer
in America," Latrobe's self-description, claimed "The
service of a republic is always a slavery of the most in-
exorable kind, under a mistress who does not give even
to her hirelings civil language."[28]

In the final analysis, technological freight is only as
good as the man who transports it. Benjamin Henry Latrobe was a very good man; he could plan, organize, and
complete large-scale projects with speed and style in a
frontier environment. Currently we are evaluating one of
his internal improvement projects as to mode and accuracy
of execution. Stephen F. Lintner, a doctoral candidate in
The Johns Hopkins University's Department of Geography
and Environmental Engineering, is using Latrobe's Susquehanna Survey map and his field and account books as the
base point for his study of the historical physical behavior
of the lower Susquehanna River (1801 to the present).[29]
Latrobe drew the Susquehanna River Survey map of 1801-02
upon completion of that unusual feat of ecological and navi-
gational engineering. The original map was burned in
Washington by the British. Latrobe redrew our copy in
1817. It is seventeen feet long and two feet wide, a
single annotated map, graphically rendered in color, in-
dicating rapids, rocks, forms, and geological elements
of the river and its neighboring ridges from Columbia to
the Maryland line.

A plant geographer, Mr. Lintner has already carried
out extensive fieldwork on a twelve-mile stretch of the
river from Columbia to Safe Harbor Dam, Lancaster County,
Pennsylvania. Employing Latrobe's map as a base, he has
created four overlay maps from later nineteenth and twen-
tieth century surveys and has completed land, river, and
aerial reconnaissance of the site. Latrobe's technical sur-

vey and biogeology have been checked by Mr. Lintner, who
has participated in historical cartographical surveys in
California. His preliminary findings are truly remarkable.
The map constitutes a high level technological/scientific
survey, presentation, and analysis of the physical and bio-
geological features of the river and its valley. The carto-
graphic work conforms closely to later and more scientific
measurement. The geology is correctly defined. The bio-
logical (flora) and economic (land use) statements are of
great historical value. The speed with which the survey
was conducted, the channel improved, and the map drawn
was amazing. Mr. Lintner doubts that a modern team the
size of Latrobe's party could match this record.

Of a more problematic nature is Benjamin Henry
Latrobe's role as an accelerator of technological momentum.
The engineer was sought after because of his ability to plan
and execute large-scale integrated technological projects
such as the Philadelphia waterworks, the U. S. Capitol,
and the Washington Navy Yard. He was an environmental-
ist in that he conceptualized relationships: man to a
structure, structures (domestic, civil, commercial, and
industrial) to an urban center, the urban center to its
immediate physical surroundings, and the locale to the
region.[30] The scale of many of his projects was com-
parable to large local European undertakings. Latrobe
often produced the entire technological package as did
British railway engineers of the next generation. Because
of the demands on him (he had projects running simulta-
neously in Washington, Baltimore, Delaware, and Philadelphia from 1803 to 1806), he attempted to develop support facilities by continuously encouraging the development
of steam engines, building techniques and materials, and
machinery of all sorts. At this juncture, one can only
speculate that all of these factors first concentrated technology, next accelerated its growth, and then facilitated
its transfer or dispersal throughout the United States.
Certainly Latrobe's major projects required multiple
technology and large capital investment which strained
available American resources.[31]

The Philadelphia waterworks created technological momentum of sorts. It was the most sophisticated engineering project of the period, attracted national attention, was fully documented, and resulted in great social benefit. Latrobe found capable contractors among whom he could apportion the work. This operation provided technical leadership, skilled labor, a pool of heavy Irish labor, and an acquaintance with American construction capabilities and costs that the engineer next utilized on the Chesapeake and Delaware Canal. So successful was he at mobilizing labor that Aaron Burr in 1805 enticed Latrobe to "make arrangements to engage and send to the Western country, 500 Irish laborers, to be employed in cutting a canal at the falls of the Ohio . . . ."[32] Luckily he hesitated and avoided direct involvement in the New Yorker's western schemes. Latrobe's practical work with steam engines began at this time. Carroll W. Pursell shows how from then on Latrobe worked closely with engine builders, continually suggesting "innovations" in engine design. [33] Most important, the Philadelphia project led to other waterworks planned and directed by "Philadelphians": New Orleans, Latrobe's undertaking initially directed by his son Henry; Baltimore, built by John Davis (1770-1864) former clerk of the works; and Fairmount (Philadelphia), executed by Latrobe's chief draftsman Frederick Graff (1774-1847).

It was as Surveyor of Public Buildings and Engineer of the Navy that Latrobe expended the greatest sums of money and encountered the widest range of challenges to his capabilities as a structural engineer. His first commission in 1802 for President Jefferson was to plan covered drydocks in which existing warships could be stored and preserved in peacetime. The construction cost of the covered dry dock and two large masonry locks was greater than Congress would appropriate--$417,267. [34] Latrobe went on to do technological work on the U. S. Capitol, the President's house, Pennsylvania Avenue, the Treasury Department's fireproof vaults, the Navy Yard, and to draw plans for the Marine and Naval hospitals. The U.S. Capitol involved, as noted above, extensive structural, acoustical, and heating engineering in both the initial and rebuilding phases. Latrobe was forced to create technology for the development and transportation of building materials. [35] This activity required the creation of a building industry in a frontier village. It set technological standards for years to come. His report and estimate, $984,852, on the New York-Long Island (Blackwell's Island) Bridge of 1804 was too grand for the economy to bear but is evidence that the engineer believed Americans could provide technological support for such a venture because of the experience gained in the Philadelphia waterworks and Schuylkill bridge projects. [36]

Between 1798 and 1820, Benjamin Henry Latrobe may well have been the single most important American agency of technological dissemination. Probably this aspect of his role as an agent of technological transfer did not quite equal his own work, but it came close to doing so. For as David Landes suggests about the transmission of British industrial technology to Europe in the first half of the nineteenth century: "Perhaps the greatest contribution of these immigrants [British technicians] was not what they did but what they taught . . . ." [37] Latrobe, something of a one-man technical university, taught, trained, promoted, researched, published, and advised. He was continually bombarded with requests for his professional advice and nominations for those posts he could not fill himself, and he traveled extensively throughout the United States.

Consider this brilliant list of his students and former subordinates, Latrobe protégés all: Robert Mills, William Strickland, Louis DeMun, Henry Latrobe, Frederick Graff, and William Davis. Among these men were some of the most important and creative engineers of the next generation; Strickland and Mills also "became in turn the country's most distinguished architects to carry forward the development of the profession . . . ." [38] Also John H. B. Latrobe served as a draftsman for the rebuilding of the Capitol and the Second Bank of the United States competition drawings, received engineering training at West Point,

became Chief Counsel of the Baltimore & Ohio Railroad
where he shaped technological policy and helped promote
his brother, Benjamin H. Latrobe, Jr., (1806-1878), to
the post of Chief Engineer. Whenever possible, the elder
Latrobe also advanced the careers of his protégés.

He published reports and research papers on a wide
range of technological topics such as steam engines, the
Philadelphia waterworks, the Chesapeake and Delaware
Canal, mining, river behavior, and acoustics.[39] Also
he was called upon by governmental bodies for advice on
internal improvements. These important state papers
analyze the developmental problems, propose plans and
methods of solution, and suggest personnel to direct the
various projects. The 1808 report to Secretary Gallatin
noted above was only a précis of their correspondence regarding an integrated national plan.[40] He also rendered
similar service to the states of Virginia (1816) and North
Carolina (1818). Latrobe's letters to President Madison
and John Spear Smith concerning the appointment of a
principal engineer to direct the "improvement" of Virginia roads and canals constitute perhaps the best statement available on the conditions and requirements of
American civil engineering.[41]

It is too early to judge the degree to which Latrobe
served as an agent of continuing technological transfer
from Europe to America. Charles E. Peterson believes
that shortly after his emigration he ceased to be "on the
leading edge of British technology" by missing the importance of iron structure, being ill disposed toward stuccos,
and not keeping in touch with the developments in hydrau-
lic cements.[42] Further research may improve Mr. Peterson's estimate for we know Latrobe endlessly searched for
functional cements.[43] The engineer throughout his American career sought to build up his professional library by
European purchase, and at least once attempted to secure
engineering works currently employed at the École Poli-
technique. However, our file on Latrobe's library does
not include orders to European booksellers for newly published technological works. He was instrumental in influ-

encing English and French engineers to emigrate to the
United States. Often these men wrote directly to Latrobe
and inquired about prospects in America; some were sent
to him by friends; and several times he instituted recruiting searches for state governments, corporations, and private citizens.

Benjamin Henry Latrobe, agent of technological transfer, had a profound impact on this nation's history. He
established professional standards where there had been
few. He executed a wide range of large-scale, sophisticated technological projects that often required heavy capital investment. Many of the best engineers of the next
generation were trained by him. The great architect and
engineer skillfully articulated the value of technological
theory and spoke for "innovation" and change. Thus in
many ways Latrobe served the cause of transfer of technology not only from across the Atlantic but also within
these United States.

FOOTNOTES

1. March 23, 1796, The Journals of Benjamin Henry Latrobe, The Papers of Benjamin Henry Latrobe, Maryland Historical Society, Baltimore, Maryland (hereinafter cited as Latrobe Journals). Talbot Hamlin, _Benjamin Henry Latrobe_ (New York: Oxford University Press, 1955), pp. 526-27. Newly discovered correspondence between his elder and younger brothers in England and Livonia throws light on Latrobe's state of mind during his final days, ". . . . from the time he went to live at New Orleans, I had great misgivings about him. It is a most pestiferous place, and thousands were carried off almost every year by the yellow fever. It is strange, that in his last letter to me he speaks most slightingly of it, and says that with very little caution, anybody may remain uninfected." Christian Ignatius Latrobe (1758-1836) to John Frederic Latrobe (1769-1845), September 18, 1821, collection of Dr. John Henry deLa-Trobe, Hamburg, Germany.

2. Brooke Hindle, _Technology in Early America: Needs and Opportunities for Study_ (Chapel Hill: University of North Carolina Press, 1966), especially pp. 17-25.

3. Daniel Hovey Calhoun, _The American Civil Engineer: Origins and Conflicts_ (Cambridge: MIT Press, 1960), p. xi.

4. Quoted in _ibid._, p. x.

5. _Ibid._, p. 5.

6. The early American practice of recruiting European technological talent is documented in such articles as Carroll W. Pursell, Jr., "Thomas Digges and William Pearce: An Example of the Transit of Technology," _William and Mary Quarterly_, 3d Ser. 21 (October 1964): 551-60 and also Norman B. Wilkinson, "Brandywine Borrowings from European Technology," _Technology and Culture_ 4 (Winter 1963): 1-13. For a fuller bibliography on the transfer of technology to America see Hindle, _Technology in Early America_, pp. 88-90.

7. I am employing the term technological momentum in a far broader sense than does Thomas P. Hughes who

examined technological acceleration within one industry
and its economic and political ramifications in his valued
article, "Technological Momentum in History: Hydrogena-
tion in Germany, 1893-1933," Past and Present 44 (August
1969):102-32.

8. A new study of Cockerell's famous architect son
will tell us something of the type of training Latrobe received. See David Watkin, C. R. Cockerell: His Life and
Work (London: Zwemmer, 1974).

9. His aversion to wing dams or dykes for clearing
river channels dated from that experience. Latrobe to
Thomas More, January 20, 1811, The Letterbooks of Benjamin Henry Latrobe, The Papers of Benjamin Henry Latrobe, Maryland Historical Society, Baltimore, Maryland
(hereafter cited as Latrobe Letterbooks).

10. Latrobe's plans of 1793 and 1794 were accepted
locally but died in Parliament because of the renewal of the
French war. Hamlin, Benjamin Henry Latrobe, pp. 47-48.

11. Conversation with Charles E. Peterson, April
24, 1974.

12. Latrobe to Mr. Chequiere, November 12, 1804,
Latrobe Letterbooks.

13. Hamlin, Benjamin Henry Latrobe, p. 160.

14. Ibid., pp. 485-86.

15. Ibid., pp. 544-65.

16. The talented English engineer, William Weston
(1752-1833), also worked in America in the 1790s and did
much to invigorate and improve canal building.

17. "Mr. Latrobe's Communication." Report of the
Secretary of the Treasury On the Subject of Public Roads
and Canals; Made in Pursuance of a Resolution of Senate of
March 2, 1807. (Washington: printed by R. C. Weightman,
1808), pp. 79-107.

18. "First Report of Benjamin Henry Latrobe, to the
American Philosophical Society, Held at Philadelphia; in
Answer to the Enquiry of the Society of Rotterdam, 'Whether
Any, and What Improvements Have Been Made in the Con-
struction of Steam-Engines in America?'" Transactions of

the American Philosophical Society, vol. 6 (1809), pp. 89-
98. The report is dated May 20, 1803 and contains one plate
of drawings.
     19. For Latrobe's interest in integrated urban and
industrial design see Edward C. Carter II, "Benjamin
Henry Latrobe and the Growth and Development of Washington, 1798-1818," Records of the Columbia Historical Society
of Washington, D. C. , ed. by Francis Coleman Rosenberger
(Charlottesville: University Press of Virginia, 1973), pp. 128-
49.
     20. Carroll W. Pursell, Jr. , Early Stationary Steam
Engines in America: A Study in the Migration of a Technology
(Washington, D. C.: Smithsonian Institution Press, 1969),
passim, quotations, pp. 36-37.
     21. Hamlin, Benjamin Henry Latrobe, p. 560.
     22. Ibid. , p. 562.
     23. Ibid. , p. 553.
     24. Ibid. , p. 554.
     25. Ibid. , p. 545.
     26. Ibid. , pp. 545-46. Latrobe strongly objected to
the use of wooden locks in the Washington City Canal, which
were, in due course, ruined by a severe storm.
     27. Calhoun, American Civil Engineer, p. 16.
     28. Latrobe to Constantin F. C. Volney, July 28, 1811,
Latrobe Letterbooks, quoted in Hamlin, Benjamin Henry Latrobe, p. 564.
     29. This is a study to determine the sediment transport
characteristics and the way in which the channel of the Susquehanna River has behaved over the years. The remarkable
map of the lower Susquehanna River by Benjamin Latrobe in
1801 provides a base point. Field studies already begun,
supplemented with historical data including maps and later
aerial photos, allow for an accurate and detailed examination of the following: (1) geological and botanical processes
by which sediment becomes stablized in the channel (prin-
cipally as large vegetated alluvial islands); (2) measurement of the magnitudes, frequencies, and rates of the processes (flooding, ice jamming, and ice gorging) which serve

to transport sediment in the channel; and (3) measurement
of the response of the channel to changes in sediment type
and supply resulting from nearly 200 years of varying land
use, including a dated period of extensive coal mining. (This
is Mr. Lintner's project précis.)

    30. For Latrobe's requirements of good city planning
see Carter, "Latrobe and the Growth and Development of
Washington, " pp. 133-36.

    31. For example, by 1806 the total cost of the Phila-
delphia waterworks was $349,016.50, which exceeded the
original estimate threefold.

    32. Latrobe to Secretary of the Navy Paul Hamilton,
December 15, 1807, Latrobe Letterbooks, quoted in Hamlin,
Benjamin Henry Latrobe, pp. 223-24.

    33. Pursell, Early Station Steam Engines, pp. 34-37.
To Latrobe, "innovation" meant "a change which, while
reasonable in theory, had not yet been shown in practice to
be an improvement. " Latrobe wanted his innovations proven
functional and then incorporated in future models. All this
may have, in fact, hindered technological momentum, but
it did restate the concept of the engineer as one who sought
continuing technological improvement within a given field.

    34. Message from the President of the United States,
Transmitting Plans and Estimates of a Drydock, for the
Preservation of Our Ships of War. (Washington City: printed
by William Duane & Son, 1802). This contains a letter from
Latrobe to Robert Smith, Secretary of the Navy, December 4,
1802; estimates by Latrobe; and an extract of a letter from
Latrobe to the President, December 15, 1802.

    35. Latrobe advised on quarrying techniques, perhaps
improved a railroad to transport stone at Acquia, and constructed the Washington City Canal, in part, to facilitate
transport of heavy materials to the west front of the Capitol.

    36. Latrobe to Nicholas Roosevelt, November 10,
1804, Latrobe Letterbooks, quoted fully in Hamlin, Benjamin
Henry Latrobe, pp. 578-82.

    37. David Landes, The Unbound Prometheus (Cam-
bridge: Cambridge University Press, 1969), p. 150 cited in

Nathan Rosenberg, "Economic Development and the Transfer of Technology: Some Historical Perspectives," Technology and Culture 11 (October 1970):555.
    38. Hamlin, Benjamin Henry Latrobe, pp. xviii-xix.
    39. For example, in addition to his report in the
American Philosophical Society Transactions cited above in
footnote 18, see View of the Practicability and Means of Sup-
plying the City of Philadelphia with Wholesome Water, in a
Letter to John Miller, Esquire, from B. Henry Latrobe,
Engineer. Dec. 29, 1798. (Philadelphia: printed by Zacha-
riah Poulson, Junior, 1799); American Copper-Mines (Philadelphia: n.p., 1800); Opinion on a Project for Removing the
Obstructions to a Ship Navigation to Georgetown, Col. (Washington: W. Cooper, printer, 1812); "Acoustics," in New
Edinburgh Encyclopaedia (Philadelphia: Parker and Dela-
plaine, 1832), I, pp. 104-24.
    40. See also "Letters [Oct. 31, 1813 and Feb. 10,
1814] to the Honorable Albert Gallatin, Secretary of the
Treasury of the U. S. Regarding the Road from Cumber-
land towards Brownsville," The Emporium of Arts and
Sciences 3 (1814), 285-97.
    41. Latrobe to President James Madison, April 8
and 16, 1816, Latrobe to John Spear Smith, April 25, 1816,
Latrobe Letterbooks. Daniel Calhoun employs the first Madison and Smith letters, noting the engineer was the Virginia
Board of Public Works' "most expatiatory informant," to
analyze the make-up of the engineering profession in 1816.
American Civil Engineer, pp. 18-20, quotation p. 18.
    42. Conversation with Charles E. Peterson, April 24,
1974.
    43. Hamlin, Benjamin Henry Latrobe, pp. 306, 366.

MONCURE ROBINSON: RAILROAD ENGINEER,
1828-1840

Darwin H. Stapleton

     Three overlapping but distinguishable phases in the
relationship of American to European civil engineering
technology occurred in the United States between about
1790 and the mid-nineteenth century.   First, the European civil engineer came to the United States to fill a
profession for which no American had the necessary skill.
Second, the American civil engineer visited Europe to
learn new skills.   The third phase was that of common in-
terchange when there was a mutual interest in technical
developments on either side of the Atlantic.
     This thesis rests on some assumptions which are
difficult to prove but nonetheless arguable.   First, I as-
sume that technology is not external to man, that it has
intangible elements which render it fully communicable
only between persons who are in each other's presence.
It is clear to me that a technical skill is usually so complex as to be insusceptible to any type of indirect communication, such as printed materials or another's ob-
servations, and must be taught directly to others by the
man who has the skill.   This is not a new idea on my part,
but it indicates why I have cast my thesis in the context of
men's activities rather than the development of machines
or other objects. [1]
     Moreover, in the historical era which I am dis-
cussing it appears that the United States was a technical
colony of Europe, specifically Western Europe, and that
the most important innovations of the period -- perhaps
I should say those which seem important to us in historical perspective -- were of European origin.   The initial
phase of the transfer of these innovations to the United
States -- that which was done by Europeans skilled in the
technology -- is probably the most familiar to students
of American history.   Samuel Slater in textiles, Benjamin Henry Latrobe in architecture and civil engineering,

and E. I. du Pont in gunpowder manufacture are some
of the better known individuals of the era.

This case study concerns the second phase of my
thesis: the visits of skilled Americans to Europe to
bring back new technical information. By skilled Americans I mean men who had learned a technical profession
or trade in the United States and had already practiced
it. Men of this type were sometimes compelled by their
own interest in European developments in their field to
cross the ocean and learn all they could about the new
technology. Moreover, skilled Europeans were usually
willing to share their knowledge, which is apparent to
anyone who has read about the European travels of
Joshua Gilpin, William Strickland, and George Escol
Sellers. [2] Here I am concerned with the experience of
Moncure Robinson, one of a small group of American
civil engineers who visited England in the 1820s and
1830s to learn about railroads. This group of men,
whom I shall mention further later in this paper, were,
I believe, responsible by themselves or through their
associates for every railroad built in the United States
before 1840.

Moncure Robinson surveyed and determined the
character of the Allegheny Portage Railroad; served
as chief engineer of the Richmond, Fredericksburg and
Potomac Railroad and other early Virginia railroads;
and superintended the construction of the Philadelphia
and Reading Railroad, to my mind one of the finest examples of civil engineering skill in early railroad history.

Moncure Robinson was born in 1802 to a merchant
family of Richmond, Virginia. His parents were concerned about his education and sent him to a tutor for
several years, until at the age of thirteen he entered
the College of William and Mary. Robinson spent three
years at William and Mary but left in 1818 before re-
ceiving his degree. [3] Sometime when he was in college
he decided to begin a career in surveying and, perhaps,

civil engineering, because in 1819 he volunteered to be
an assistant, without pay, on a state expedition to ex-
plore the possibility of constructing a canal to connect
the James and Kanawha rivers. This arduous service
of about seven months in the rugged mountains of what
is now West Virginia earned him special notice in the
expedition's official report to the Virginia Assembly. [4]
     Less than three years later, in the summer of
1822, he went on a personal journey. He traveled to
New York to examine the Erie Canal, then still under
construction, probably in hopes of becoming better
qualified for employment by his own state's James
River (canal) Company. He saw the full length of the
Erie, made drawings, and talked with canal engineers.
Robinson told his father:

> ... I have met with nothing but attention
> and civility whilst engaged in my re-
> searches on the canal & have found no
> difficulty in obtaining all the information
> which I wished -- so that I now feel (if
> I do not flatter myself) prepared to exe-
> cute any work for which we shall have
> occasion on the James River. [5]

Within a year after returning to Virginia he was
hired by the state as engineer for the construction of
an extension of the canal upriver from Richmond. For
two years he filled that position, surveying the route
to be followed, being responsible for the contractors
who were doing the construction, and generally carrying out the orders of the state commissioner to whom
he was responsible. [6]   He did his work well according
to an official report and did not leave the James River
canal until the section he was working on was nearly
completed. [7]
     He left for one negative reason: it did not look
as if Virginia was likely to support further work on the

James River canal or any other internal improvement;
and one positive reason: he was enthralled by the possibility of visiting Europe. [8] Undoubtedly a great part
of his enthusiasm for the trip came from a desire for
adventure, but he certainly had professional objectives
in mind when he wrote to a friend who was already in
Europe that:

> . . . Instead . . . of Stratford on Avon I
> should make a pilgrimage . . . to the works
> of the Duke of Bridgewater . . . . Instead
> of an excursion in a steam-boat to the Isle
> of Man I shall probably make one in a post-
> coach to the factory of those great genera-
> tors of <u>steam</u> & <u>Steam Engines</u> Messrs.
> Bolton & Watt . . . . [9]

It was with that attitude that Robinson made his decision
to go to Europe.   It was an important decision both for
his own life and the history of technology in the United
States, as he was not only one of the first American
civil engineers to seek to acquire European skills, but
he was also one of the first Americans to see the beginning of the modern railroad age in England. [10]
     He crossed the ocean in the spring of 1825 and re-
mained in Europe for two-and-a-half years.   Although
he was in England for at least seven months, most of
the time he lived in Paris, studying at the Sorbonne for
two winter sessions (1825-26 and 1826-27).   He attended
lectures given by the physicist and chemist Gay-Lussac
and the chemist Thenard. [11] It is not clear what subjects
they were teaching, although Robinson referred to his
Parisian studies as if they concerned his "profession"
or the "theory" of civil engineering. [12] That seems reasonable because both Gay-Lussac and Thenard were
trained at the École Polytechnique, which was the leading technical school in Europe at the time. France as
a whole was the center of technical education in the

world.[13] For example, in civil engineering the famous
lectures of Louis Navier which established structural analysis and   strengths of materials theory as we know them
today were first published in 1826.[14] The Parisian world
in which Robinson was a student was an extremely fertile
one for his profession.

His pursuit of knowledge was not completely theoretical, however. He investigated machines, factories,
bridges, aqueducts, and a hundred other things of interest. Initially his primary object was to see canals. A
friend with whom he was traveling in France found that
Robinson's desire to observe them in every locality could
not be dampened.[15]   After going to England with the same
friend he went off alone walking the towpaths of the Grand
Junction and other canals to satisfy his curiosity.[16]

The first mention of railroads in any of Robinson's
correspondence occurs during his trip through England
in the summer and fall of 1825. English railroads immediately made a strong impression on the young American, and he appears to have grasped their significance
rather quickly. A letter of his which was published in
the Richmond _Enquirer_ late in 1825 included his proposal
to use them to meet the transportation needs of Virginia.
He suggested that the state build 130 miles of railroad to
connect canals built to the upper reaches of the James and
Kanawha rivers to make a through transportation route
from the Ohio River to tidewater in Virginia. He noted
that "steam waggons," or locomotives, could pull three
or four carriages on a railway and that inclined planes
could overcome any steep grades that were encountered.[17]
Although enthusiasm may be too strong a word to describe
Robinson's reaction to railroads, I think it fair to say that
he immediately understood them as a feasible means of
transportation for the United States. He observed and investigated railroads again for several months in 1827 just
before returning home.[18]

Robinson's letter was not the first news of railroads
to cross the Atlantic, although it was near the very begin-

ning of serious consideration of railroads in the United
States.   Late in 1824 a Baltimore merchant, possibly an
agent of Alexander Brown & Sons, visited England to
collect information on railroads. [19] Shortly thereafter
the Pennsylvania Society for the Promotion of Internal
Improvement was founded in Philadelphia, and early in
1825 it sent William Strickland, a student of Benjamin
Henry Latrobe, to Europe as its agent.   The reports
which he sent back were full of information about railways, and he encouraged their construction where traffic was heavy and speed important. [20]   A digest of a
Robert Stephenson essay as well as a pamphlet by the
respected American editor and economist Mathew Carey
were in circulation. [21]   With the information they had
on hand in 1825 Americans were already involved in
debates over whether to build railroads.   Over the next
couple of years an American who read a newspaper was
well aware of railroad developments in England.

        When Robinson returned from Europe he found the
usefulness of railroads at least partially accepted here,
but unfortunately for him Virginia was one of the places
where the idea had not taken root.   The rapidly expand-
ing Pennsylvania state works attracted his attention, and
perhaps with the help of some Philadelphians he had met
in Paris, he received an appointment from the Pennsylvania Canal Commissioners in the spring of 1828. [22]   The
Commissioners recognized that he had a special understanding of railroads and assigned him to survey possible
routes between the North Branch of the Susquehanna and
the upper reaches of the Schuylkill and Lehigh rivers,
both tributaries of the Delaware.   That he was willing to
take such a position is an indication of his positive as-
sessment of the future of railroads.   Instead of canals,
to which the recently completed Erie Canal had attracted
Americans' attention, Robinson chose to associate himself with railroads, which had not yet appeared in modern
form here.
        While in England Robinson undoubtedly saw the

Stockton & Darlington, the first modern railroad with
locomotives and steam-powered inclined planes. The
Stockton & Darlington was put into operation in 1825,
and the engineer of that road, George Stephenson, then
moved to the Liverpool and Manchester which was still
under construction in 1828. The famous Rainhill trials,
which were made to test the potential of locomotives on
that road and which drew international attention to it,
took place in the fall of 1829. [23]

The United States, not having had experience
with tramways, was much more hesitant than England
in taking its first steps into the railroad era. Its first
two railroads, the Granite Railroad in Massachusetts
and the Mauch Chunk railroad in Pennsylvania, did not
have locomotives or steam-powered inclined planes and
were built by self-taught American engineers. The Gravity Railroad in northeastern Pennsylvania was just beginning construction in 1828. When completed a year
later the road used steam engines on its inclined planes
and attempted to use locomotives. One of the two engineers on the Gravity, Horatio Allen, visited England
in 1828 to see and learn about railroads, especially the
Liverpool and Manchester. [24] The most widely publi-
cized early American railroad was the Baltimore and
Ohio, which was chartered in 1827 but did not have its
first segment of thirteen miles opened until 1830. The
promoters of this railroad clearly intended to follow the
principles of the English roads. They sent three engineers (Jonathan Knight, George Whistler, and William
Gibbs McNeill) to England late in 1828, and they returned in May, 1829, full of confidence in the practicality of locomotives. They then laid out the B & O so it
was suitable for locomotives and planned an inclined
plane to surmount the first ridge. [25]

It is my impression that every American railroad
built in the first decade after the opening of the B & O
had associated with it an American civil engineer who
had visited Europe or who had worked with a visitor.

The list of these visitors includes William Strickland,
Horatio Allen, and the B & O engineers already mentioned, as well as Samuel Kneass, engineer of part of
the Philadelphia, Wilmington and Baltimore Railroad,
Edward Miller, who supervised the construction of the
inclined planes on the Allegheny Portage, John Edgar
Thomson, who became the Chief Engineer of the Pennsylvania Railroad, and several other major civil engineers.[26] The number of men whom they taught through
the on-the-job training system prevailing in civil engineering at this time is difficult to estimate but certainly
was large.[27]

There were a few other railroads in some stage
of construction in 1828, and many others projected,
but the knowledge of railroads was in its infancy when
Robinson accepted the challenge of surveying railroad
routes for the Pennsylvania Canal Commissioners.[28]
With a party of several assistants he carried out care-
ful surveys over the rugged Appalachian terrain of what
is now Schuylkill, Carbon, and Luzerne counties during
the spring, summer and fall of 1828.[29]

Robinson's report on the Susquehanna-Delaware
surveys shows the English basis of his engineering
principles. He noted that many believed that American
railroads could be cheaper to construct than those in
England because wood could be substituted for stone in
the roadbed (i.e., wood instead of stone sleepers), and
wood rails plated with iron could be used instead of cast
iron rails. To this change of materials in America he
had no objection, but he pointed out that no alteration was
likely to be possible with respect to English principles
for railroad grades. One of those principles was that
it was necessary to plan the grade which a road would
have with the direction of trade in mind. If equal trade
was expected in both directions the road should be as
level as possible throughout. If trade was expected in
primarily one direction the major restriction was that
the road could not be so steep in the return direction so

as to hinder the return of empty cars by the motive power.
He quoted results from experiments on the Stockton &
Darlington Railroad which indicated that this graduation
could not exceed 18 feet to the mile (0. 35% or about 0. 6°).

With these principles restricting the grade of railroads, there was little difference between the routes
which they could take and the service they could provide
and that of canals or turnpikes, Robinson noted. He was
referring to the fact that at this date railroads as well as
canals and turnpikes depended largely on horses for motive power. But what had radically changed the applica-
bility of railroads in recent years was locomotives which
increased the speed of travel, and inclined planes with
stationary engines, which allowed railroads to climb
grades previously impossible. He believed that these
two innovations vastly increased their capabilities. [30]
Robinson's recommendation of possible railroad routes
was therefore based on four principles: (1) the routes
should cross the mountains at the least possible eleva-
tion; (2). there should be the gentlest grade possible in
the direction of most trade, which he expected in this
case would be from the Susquehanna to the south; (3)
routes should be sought which could be used by locomotives; and (4) intolerable grades should be overcome
with inclined planes. [31]   There were echoes of these
English ideas throughout Robinson's professional career.

Only three days after the Canal Commissioners had
received and read his first report on the Susquehanna-
Delaware surveys they appointed him engineer on the
Allegheny Portage. [32]   Their resolution accompanying
the appointment asked Robinson to survey the route and
then give his opinion on whether a railroad, turnpike or
"any other plan" would be the best means of transportation. [33]   Behind these words was a political turmoil of
more than three years which Julius Rubin has illumina-
ted in his book <u>Canal or Railroad?</u>   With the first wave
of enthusiasm over internal improvements in 1825, a
canal had been planned from Philadelphia to Pittsburgh

which needed a four-and-a-half mile tunnel through the
Allegheny summit. Arguments over feasibility and ex-
pense prevented the tunnel from being started, although
the canal sections in the Allegheny and Susquehanna
river systems were begun by the state. The two sections were useless for east-west trade without some
sort of mountain connection, however. By the time
Robinson was appointed to give his opinion about what
could be done, there was widespread debate over whether
to build a railroad, canal or turnpike and what route to
follow. [34] I think it is some indication of the value which
was placed on his knowledge that he was given the responsibility for this sensitive decision.

There is no evidence that he hesitated to accept the
responsibility; to the contrary, he was exhilarated by
the challenge. In a letter to his father he explained:

> . . . I am to push my explorations into
> every glen and nook of the Allegheny
> through which it may be possible to creep
> with a railroad. Should I succeed in find-
> ing a route favorable enough to be adopted
> and executed I shall be much pleased, as
> it will afford me an opportunity of making
> myself known in a department of my pro-
> fession, in which few engineers have as
> yet embarked. [35]

This was an opportunity which might push him to the top
of his profession: if his opinion was accepted he might
be chosen engineer in charge of construction and be identified with one of the most important railroad projects of
the time. [36]

He worked on the Allegheny Portage survey with
six assistants from April until August of 1829. [37] In
his report filed three months later he reiterated the
principles he expressed in the Susquehanna-Delaware
surveys and based his recommendations on them. He

suggested a railroad using locomotive power on level
stretches, water and steam power on several inclined
planes which would overcome steep grades, and a one-
mile tunnel to reduce the height of the summit. [38] Unfortunately, this survey was as difficult for the Pennsylvania legislature to accept as that which had proposed
the four-and-a-half mile canal tunnel. No action was
taken on it, and in March, 1830, another survey of the
area was authorized. [39]

As it turned out, the Allegheny Portage Railroad
was built essentially as Robinson had envisioned it, but
it may have been the Pennsylvania legislature's rejection of his recommendations that made him turn completely away from state to private enterprise. That
change had been brewing for some time, as he was already involved with some railroad promotional activities in Virginia and Pennsylvania early in 1829. [40] In
addition, it must have been clear to him that state
governments were not involved in railroad construction
to the extent that private corporations were. At any
rate, from the late summer of 1829 Robinson was always connected with nonpublic railroad projects and
only occasionally performed professional services for
a state. His career became considerably more compli-
cated, as he was often associated with several projects
at the same time, a pattern typical of American civil
engineers of this era.

It is difficult to follow his career chronologically
from this point, just two years after he had returned
from Europe and when he was 27 years old. Yet it is
interesting to look into the different tasks which he
performed because they demonstrate the wide variety
of ways in which Robinson's skills affected early railroad development in the United States. His influence
may be gauged by one author's estimate that by 1840
Robinson was involved in some capacity with over one-
third of the railroad mileage in the United States. [41] I
have summarized his work under the headings of "consultation, " "surveying, " and "construction. "

The least demanding service which was requested of him was consultation. He was sought for consultation because his ideas were highly regarded and sometimes because an impartial opinion was needed. Such was the case when he was called to advise about a politically contested survey, or even to conduct a resurvey. The Pennsylvania Canal Commissioners employed him in that capacity twice, once in 1829 when there was a dispute over the termination of the Columbia and Philadelphia Railroad at Philadelphia, and, almost incredibly, again in 1830 when he was appointed to a board of three engineers to determine the validity of his own Allegheny Portage survey![42] Several years later with Jonathan Knight and Benjamin Wright he formed a committee to consider the plan of the New York and Erie Railroad and determine its practicality.[43] In the late 1830s and early 1840s, he was asked by several European governments to give advice on railroad construction. Russian officials apparently asked him to emigrate and help with their railroad programs, but he refused.[44] Having gone to Europe in 1825 ignorant of railroad technology, less than fifteen years later he had become recognized as one of the great railroad engineers of the world.

Surveying was the skill which he most frequently exercised during his career. This task required that he determine a feasible route for the railroad and then give a report to the persons who had commissioned him. He did at least fifteen of these surveys, and most of the proposed railroads were actually built. His surveys were closely related to railroad promotion.[45] He usually included a promotional statement in his reports which suggested that the proposed railroad would carry extensive traffic and would prove profitable for the stockholders.[46] The reason for this promotional activity is clear: if the survey showed that the railroad traversed a feasible route and could make money, enough capital might be gathered to build it. In every case but one, the railroads for which Robinson was chief engineer during construction also had the original survey done by him.[47] As his career advanced

he was less willing to take on surveying duties, and in
1838 it took the offer of compensation at the rate of $10,000
per annum, well over his usual fee, to induce him to survey the proposed Brunswick and Florida Railroad.[48]

The reason for Robinson's reluctance was his increasingly heavy responsibilities as chief engineer in
charge of the construction of railroads. In his civil engineering career he directed the construction of seven
railroads and was always the chief engineer of two or more
at the same time.[49] The first two of these railroads,
built from 1830 to 1831, were the Chesterfield in Virginia
near Richmond and the Little Schuylkill in Schuylkill
County, Pennsylvania. Both were coal feeders; that is,
they ran from coal mines to water transportation and had
continuous downgrades in the direction of trade. Typical
of this early stage of American railroads, they were built
to use horses for motive power, although Robinson insisted
on building the Little Schuylkill so that it could use locomotives. Two English ones were imported in 1833 and
successfully used on it.[50]

The Petersburg Railroad, a sixty-mile route from
Petersburg, Virginia, to Weldon, North Carolina, was the
first road on which Robinson was able to fully implement
his English railroad principles. Built from 1831 to 1833,
it was laid out with a very even and low grade and had an
inclined plane at Weldon. English locomotives were used
on the Petersburg from its opening.[51] Concurrently with
the Petersburg's construction he was engineer on the Danville and Pottsville Railroad which ran between those two
Pennsylvania towns. Challenged by the Appalachian ridges which had to be overcome, Robinson put in seven inclined planes, a tunnel and extensive cuts and fills to
maintain the lowest possible grade for traffic. An interesting aspect of the Danville and Pottsville was the use of
water at the top of the planes to fill specially designed tank
cars for use as counterbalances to cars loaded with coal.
This tactic eliminated the need for stationary engines at all
but one of the planes.[52]

The last three railroads which Robinson built represent his maturity as an engineer. For the Philadelphia and Reading he laid down three principles which are clearly reminiscent of those which he had enunciated in 1828 based on his English experience: (1) no grade in the direction of trade should be ascending; (2) no other grade should exceed eighteen feet to the mile; and (3) the shortest radius of curvature should be 818 feet. [53]   The narrow, sinuous valley of the Schuylkill which this railroad followed, and which already contained turnpikes and a canal, was not the easiest situation in which to adhere to any of these principles. The route as completed from Philadelphia to Mount Carbon in January, 1842, was double-tracked with several bridges and tunnels and was designed to accommodate 75,000 tons of anthracite per day. Richard Boyse Osborn, one of Robinson's assistants and afterward his biographer, called the Philadelphia and Reading "the crowning achievement of [Moncure Robinson's] professional career. "[54]

That distinction should probably be shared with another of these three railroads, the Richmond, Fredericksburg and Potomac, or R, F & P, for which he had made a preliminary survey in 1834. After appointment as chief engineer he laid out a route suitable for locomotives and had the entire length open from Richmond to Fredericksburg early in 1837. [55] This fifty-mile railroad cut through a rolling country crossed by several rivers and was destined to become a major link in the rail network of the eastern seaboard.

Robinson was the logical choice as chief engineer of the Richmond and Petersburg Railroad, a short but important line which connected his other major efforts in Virginia, the Petersburg and the R, F & P. In 1838, about two years after his appointment as chief engineer, he had the Richmond and Petersburg open. His bridge over the James River on that road was regarded by Michel Chevalier, a French engineer, as a major engineering achievement. [56]

Moncure Robinson, from the portrait by Thomas Sully.
Reprinted from Nathalie Robinson Boyer, <u>A Virginia
Gentleman and his Family</u> (Philadelphia, 1934), frontispiece.

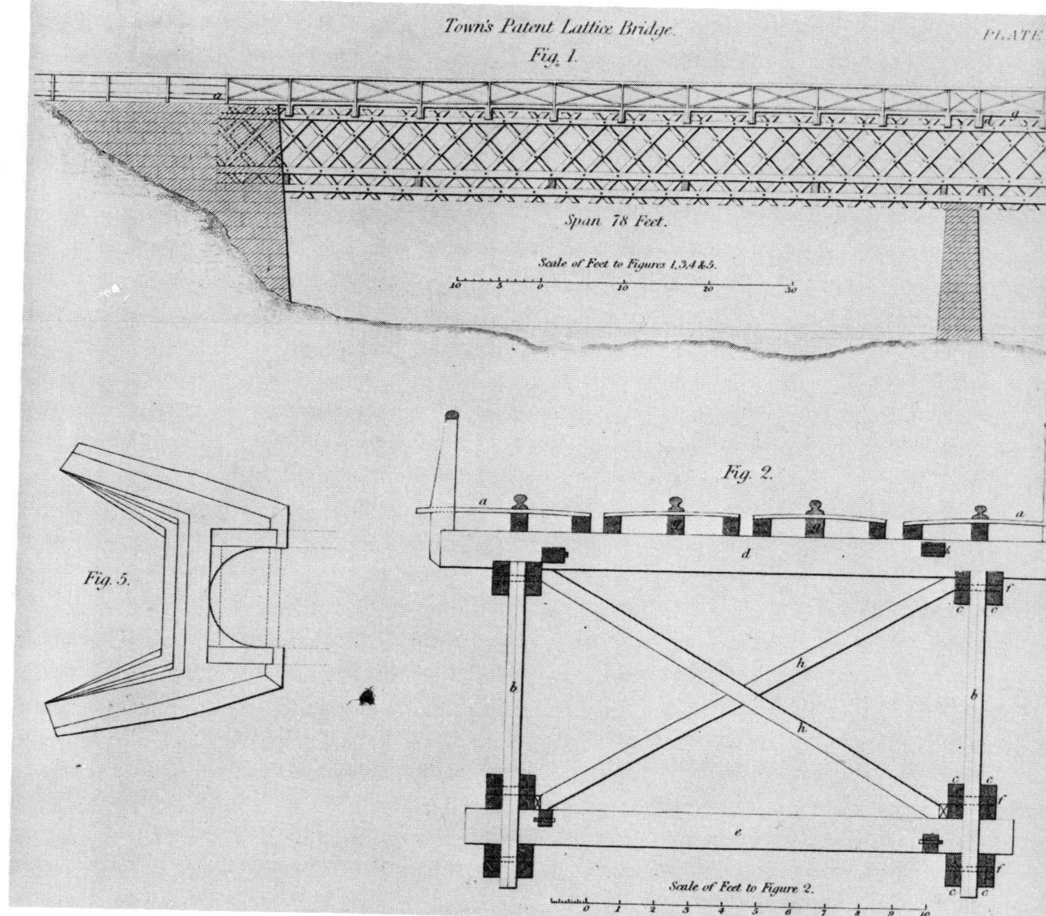

Bridge design for Philadelphia and Reading and Richmond and Petersburg railroads by Moncure Robinson. Based on the plan of Itaiel Town's wooden lattice truss. Reprinted from David Stevenson, <u>Sketch of the Civil Engineering of North America</u> (London, 1838), p. 231.

Overlapping with his duties as construction engineer were his functions as an administrator. More and more, especially with the Philadelphia and Reading and the R, F & P, Robinson spent the majority of his time in the office rather than in the field. He relied upon his principal assistants to direct the field work and he attended to problems of organization and finance.[57] Accompanying this change of function was his heavy investment in the Virginia railroads to the extent that he became a large private stockholder.[58] The culmination of that involvement was his election in 1840 to the presidency of the Richmond, Fredericksburg and Potomac Railroad and the effective end of his civil engineering career.[59] Robinson was one of the first civil engineers to make the change from engineering to administration, a path later followed by J. Edgar Thomson and John B. Jervis, among others.[60] As Eugene Ferguson has pointed out to me, the move to administration has been an indicator of an engineer's "success" ever since. That is, it usually elevated a man's social status and frequently his financial status as well. This phase of his career lasted much longer than that as a civil engineer since he lived until 1891.

In the context of Robinson's civil engineering works which have just been outlined it may be seen how his skills, based on English principles, were diffused through the United States and embedded in its railroad technology. (Of course, the same process occurred for the other American engineers who had visited England to learn about railroads.) The corps of assistants which he always had on hand was the major means of diffusion. That group was not constant; some men remained with him for only a few months, but several were with him for years. His major assistant from his earliest surveys for the Pennsylvania Canal Commissioners was Wirt Robinson, a cousin. Robinson trained him in the profession, supported him when he made his own pilgrimage to Britain in 1831-1833, and eventually worked with him as an equal in the later stages of the Philadelphia and Reading's construction.[61] Walter Gwynn was one of his associates on the early Virginia railroads, but Gwynn broke away to become

the chief engineer of the Portsmouth & Roanoke Railroad
and other southern railroads. [62] John H. Hopkins, another
Virginia assistant, became the chief engineer of the Louisa
Railroad, the oldest section of what became the Chesapeake
and Ohio. [63] Two of the men who worked with Robinson on
the Philadelphia and Reading probably had the most influen-
tial careers afterward. Richard Boyse Osborne was with
that railroad until 1845 when he went to Ireland for five
years as chief engineer of the Waterford & Limerick Railway. He returned to the United States and was involved in
a number of railroads including the Camden & Atlantic, the
Lebanon Valley, and the Western Maryland. [64] William
Hasell Wilson went on to the Gettysburg Extension and later
had a long career with the Pennsylvania Railroad. [65]

I think that the story of Moncure Robinson's transfer
of railroad technology to the United States supports my
original contentions about the transfer of technology. He
was one of a number of railroad engineers who personally
brought elements of the English experience with railroads
to this country. It flourished here because public opinion
was prepared and economic conditions were ripe. At the
same time it is clear that Robinson did not go to Europe
out of social or economic pressure for railroad engineers,
but out of his own inquisitiveness--in fact, when he went he
was concerned only with canals. This circumstance indicates both the necessity for looking at technology as having
a character of its own and the inadequacy of the easy as-
sumption that social or economic "demand" calls all innovations into being. [66] The common technological heri-
tage of the United States and Europe allowed a relatively
rapid and successful transfer of railroad technology, and,
in fact, Moncure Robinson became recognized as one of
the world's leading railroad engineers. Our increasing
knowledge of the characteristics of the transfer of technology should further emphasize the intricate international
and intercultural developments which are associated with
many technological innovations.

FOOTNOTES

<u>Key to Abbreviations of Manuscript Sources</u>

HSP   Historical Society of Pennsylvania, Philadelphia, Pa.

MRP   Moncure Robinson Papers, Earl Gregg Swem Li-
          brary, The College of William and Mary in Vir-
          ginia, Williamsburg, Va. (Many of Moncure
          Robinson's letters to his father, 1822-1833, were
          published in the <u>William and Mary Quarterly</u>, 2nd
          Ser. VIII (1928):71-95, 143-56; IX (1929):13-33.)

PaCC   RG 17, Records of Land Office, [Pennsylvania]
          Board of Canal Commissioners, William Penn
          Archives, Harrisburg, Pa.

RC   Treasurer's Vault, Reading Company, Reading
          Terminal, Philadelphia, Pa.

VaBPW   Records of the [Virginia] Board of Public Works,
          Virginia State Library, Richmond, Va.

VHS   Virginia Historical Society, Richmond, Virginia.

     1. Cf. Ingvar Svennilson, "Technical Assistance: The
Transfer of Industrial 'Know-How' to Non-Industrialized
Countries," <u>Economic Development with Special Reference
to East Asia</u>, ed. Kenneth Berrill (London: Macmillan & Co. ,
Ltd. , 1965), pp. 405-08; Nathan Rosenberg, "Economic
Development and the Transfer of Technology: Some Histori-
cal Perspectives," <u>Technology and Culture</u> XI (1970):553-54.
     2. Harold B. Hancock and Norman B. Wilkinson,
"Joshua Gilpin, an American Manufacturer in England and
Wales, 1795-1801," <u>Newcomen Society Transactions</u>, XXXII
(1959-1960):15-28, XXXIII (1960-1961):57-66;   William
Strickland, <u>Reports on Canals, Railways, Roads and Other
Subjects</u> . . . (Philadelphia: Pennsylvania Society for the
Promotion of Internal Improvement, 1826); Eugene S. Fer-

guson, ed. , <u>Early Engineering Reminiscences (1815-1840)</u>
<u>of George Escol Sellers</u> (Washington, D. C.: Smithsonian
Institution, 1965), pp. 108-34.

    3. Receipts of 1810 and 1815 for tuition, box 1, MRP;
John Robinson to Moncure Robinson, 18 December 1815,
MRP; Martha H. McGill to John Robinson, 10 November
1816, MRP; Moncure Robinson to John Robinson, 27 Feb-
ruary 1818, MRP; J. Aug. Smith to John Robinson, 4
March 1818, MRP.

    4. "Report of the Principal Engineer . . . ," in <u>Fourth</u>
<u>Annual Report of the Board of Public Works</u> (Richmond:
Thomas Ritchie, 1820), pp. 113-14; receipts for expenses
of Moncure Robinson, 1819, in (uncatalogued) Moncure
Robinson box, VaBPW.

    5. Moncure Robinson to John Robinson, 13 July
1822, MRP.

    6. Receipts for Moncure Robinson's salary, 1824-
1825, in (uncatalogued) Moncure Robinson box, VaBPW;
Moncure Robinson to Randolf Harrison, 19 January 1825,
box 1, James River Company, VaBPW; Moncure Robinson
to John Jacquelin Ambler, 11 March 1824, box 8, Ambler
Family Papers, Alderman Library, University of Virginia,
Charlottesville, Va. (hereafter cited as Ambler Family
Papers); "Report of the Commissioner of the James and
Jackson's River Canal . . . ," in <u>Ninth Annual Report. . .</u>
<u>Board of Public Works</u> (Richmond: T. W. White, 1825),
pp. 144-49; "Report of the Commissioner of the James
and Jackson's River Canal," in <u>Tenth Annual Report. . .</u>
<u>Board of Public Works</u> (Richmond: Shepherd & Pollard,
1826), pp. 304-05.

    7. "Report of the Commissioner of the James and
Jackson's River Canal. . . ," in <u>Ninth Annual Report</u>, p.
144; "Report of the Commissioner of the James and Jackson's River Canal. . . ," in <u>Tenth Annual Report</u>, p. 301;
Moncure Robinson to John Jacquelin Ambler, 21 March
1824, box 8, Ambler Family Papers.

    8. Cf. Wayland Fuller Dunaway, <u>History of the</u>
<u>James River and Kanawha Company</u> (New York: Columbia
University, 1922), pp. 73, 87. Moncure Robinson fre-

quently complained of his state's lethargy in internal improvements, e. g. , Moncure Robinson to John Robinson,
31 July 1822 and 10 February 1827, MRP.

9. Moncure Robinson to John Jacquelin Ambler,
19 July 1824, box 7, Ambler Family Papers.

10. As preceeding civil engineer visitors I am aware
of Loammi Baldwin, Jr. , in 1807-1808 and 1823-1824, Canvass White in 1817, and William Strickland (by a couple of
months) in 1825.

11. Moncure Robinson to Conway Robinson, 19 Feb-
ruary 1825 [misdated for 1826], MRP; Moncure Robinson
to John Robinson, 19 January 1827, MRP; entry for 24
January 1826, John Jacquelin Ambler journal, box 2, Am-
bler Family Papers; M. P. Crosland, "Joseph Louis Gay-
Lussac," Dictionary of Scientific Biography (New York:
Charles Scribner's Sons, 1972), V:317-27; "Louis Jacques
Thenard," in Biographisch-Literarishes Handwörterbuch
..., ed. J. C. Poggendorf, (Leipzig: Johann Ambrosius
Barth, 1863), II:1088-89.

12. Moncure Robinson to John Robinson, 10 February
1826, 19 January 1827, and 10 February 1827, MRP.

13. Frederick B. Artz, The Development of Technical Education in France, 1500-1850 (Cambridge, Mass.:
SHOT and M. I. T. Press, 1966),pp. 230-266.

14. Hans Straub, A History of Civil Engineering,
trans. Erwin Rockwell (Cambridge, Mass.: M. I. T. Press,
1964), p. 153.

15. Entry for 26 June 1825, Ambler journal, box 2,
Ambler Family Papers.

16. Entries for 5 August 1825 and 23 August 1825,
Ambler journal, box 2, Ambler Family Papers; [Moncure
Robinson] to [Randolf Harrison], 16 August 1825, in the
Richmond Enquirer, 25 December 1825.

17. [Moncure Robinson] to [Randolf Harrison], 16
August 1825, in the Richmond Enquirer, 25 December 1825;
cf. John Robinson to Moncure Robinson, 2 January 1826,
MRP.

18. Moncure Robinson to John Robinson, 4 September
1827, MRP.

19. Robert E. Carlson, "British Railroads and
Engineers and the Beginnings of American Railroad Development," Business History Review XXXIV (1960):138.
20. Ibid., 139-41.
21. Julius Rubin, Canal or Railroad? (Philadelphia:
American Philosophical Society, 1961), pp. 22-27.
22. Moncure Robinson to "parents," 1 December
[1827], MRP; Jos. M. McIlvaine to Moncure Robinson, 1
April 1828, MRP. The friends were Nathaniel Chauncey
and Henry Seybert. See: Moncure Robinson, "Obituary
Notice of Henry Seybert," Proceedings of the American
Philosophical Society XXI (1883):248-49; Elihu Chauncey
to Nathaniel Chauncey, 30 April 1826, box 6, Chauncey
Family Papers, Yale University Library, New Haven,
Conn.
23. C. von Oeynhausen and H. von Dechen, Rail-
ways in England: 1826 and 1827, ed. Charles E. Lee and
K. R. Gilbert, trans. E. A. Forward (Cambridge: The
Newcomen Society, 1971); Robert E. Carlson, The
Liverpool and Manchester Railway Project (Newton Abbott:
David & Charles, 1969); L. T. C. Rolt, George and Robert
Stephenson: The Railway Revolution (London: Green and
Co., Ltd., 1960); Moncure Robinson to John Robinson,
4 September 1827, MRP. Robinson wrote from Sunder-
land-on-Wear and stated his intent to visit the collieries
of the area. It seems likely that he saw the Stockton &
Darlington at this time, if he had not seen it in 1825.
24. Niles' Weekly Register (Baltimore, Md.), 19
May 1827, 9 June 1827, 23 August 1828; George Rogers
Taylor, The Transportation Revolution, 1815-1860 (New
York: Holt, Rinehart and Winston, 1951), pp. 76-77; Rubin,
Canal or Railroad?, pp. 83,87; "Gridley Bryant," in
Charles B. Stuart, Lives and Works of Civil and Military
Engineers (New York: D. Van Nostrand, 1871); A History
of the Lehigh Coal and Navigation Company (Philadelphia:
William S. Young, 1840), p. 18.
Erskine Hazard, Josiah White's partner in building
the Mauch Chunk railroad, may have been in England and
Wales in 1826 to learn about railroads. I have found only

one reference to the visit, but by a civil engineer who
worked on the Mauch Chunk during its construction. See:
W. Milnor Roberts, "Reminiscences and Experiences of
Early Engineering Operations on Railroads . . .," Transactions of the American Society of Civil Engineers VII
(1878):197.

On the Gravity Railroad: "Diary of Horatio Allen:
1828 (England)," Bulletin, Railway & Locomotive Historical Society 89 (1953): 97-138; Neal Fitzsimons, ed. , The
Reminiscences of John B. Jervis (Syracuse, N. Y.: Syracuse University Press, 1971), pp. 66-78; Carlson, "British Railroads . . .," pp. 143-47.

25. Carlson, "British Railroads . . .," pp. 147-
49; Rubin, Canal or Railroad?, pp. 69-77; John F. Stover,
American Railroads (Chicago: University of Chicago Press,
1961), pp. 13-14. The inclined plane at Parr's Ridge was
never supplied with a stationary engine, although one was
planned: Eleventh Annual Report . . . Baltimore and Ohio
Rail Road Company (Baltimore: Lucas & Deaver, 1837),
p. 19. On the B & O's reliance on the L & M model, see:
Jonathan Knight to P. E. Thomas, 17 February 1832 in
American Railroad Journal (New York), 7 April 1832;
entry for 14 November 1831, W. D. Lewis diary, Historical Society of Delaware, Wilmington, Del. ("P. E. Thomas
told me the width of track of the Balt & Ohio Road was the
same as that of the Liverpool & Manchester! -- that was
their pride -- the Engineers bro't out the facts from England."); W. Hasell Wilson, Reminiscences of a Railroad
Engineer (Philadelphia: Railway World Publishing Company,
1896), p. 13.

26. Carlson, "British Railroads . . .," pp. 139-49;
Solomon W. Roberts, "Obituary Notice of Edward Miller,"
Proceedings of the American Philosophical Society XII
(1871-1872):581-86; "Samuel Honeyman Kneass," Diction-
ary of American Biography, Dumas Malone, ed. , 20 vols.
(New York: Charles Scribner's Sons, 1927-1936) V, 454-55;
"John Edgar Thomson," DAB, IX, 486-87; George W. Cul-
lum, Biographical Register . . . U. S. Military Academy

(Boston: Houghton, Mifflin and Company, 1891), I: 238-39.
There are numerous references to Wirt Robinson assist-
ing Moncure Robinson, the earliest of which is in 1828:
Moncure Robinson to Joseph McIlvaine, 30 July 1828, box
1, Surveys & Corr. , PaCC.

27.   Daniel Hovey Calhoun, The American Civil
Engineer (Cambridge, Mass.: The Technology Press,
1960), pp. 47-53.

28.   Someone has observed that if all the railroads
chartered by state legislatures up to 1830 had been built
there would have been quite enough work for the first generation of railroad engineers. But the U. S. Dept. of Com-
merce, Historical Statistics (Washington, D. C. : Govern-
ment Printing Office, 1960), p. 427, lists only 23 miles of
completed railroad by the end of 1830.

29.   Moncure Robinson to John Robinson, 29 June
1828, 30 July 1828, 26 October 1828, MRP; Moncure
Robinson's letters to Joseph McIlvaine, box 1, Surveys
& Corr. , PaCC.

30.   Moncure Robinson, "Report on the survey of
Canal and rail-way routes between the waters of the Delaware and Susquehanna (4 December 1828), " copy in
Register of Pennsylvania (Philadelphia, Pa.) III (1829):
54-55.

31.   Ibid. , 56-58.

32.   Entry for 8 December 1828, box 1, Minute Books
and Indexes, PaCC.

33.   Ibid.

34.   Rubin, Canal or Railroad?, pp. 17-57.

35.   Moncure Robinson to John Robinson, 6 April
1829, MRP.

36.   Moncure Robinson to John Robinson, 18 August
1829, MRP.

37.   "List of Persons, Employed by the Canal Com-
missioners, May 30, 1829, " item for Allegheny Portage
& Rail Road, box 2, General Correspondence and Surveys,
PaCC.

38.   "Report of Moncure Robinson, Principal Engineer upon the Allegheny Portage (21 November 1829), "

<u>Pennsylvania House Journal</u>, 1827-1828, II, doc. 138, re-
printed in <u>Transactions of the American Society of Civil
Engineers</u> XV (1886):183-202.
    39. Rubin, <u>Canal or Railroad?</u>, p. 58.
    40. Cf. Moncure Robinson to John White, 5 April
1829, Adelman Collection, HSP; Conway Robinson to Moncure Robinson, 21 March 1829, MRP.
    41. Revelle Wilson Brown, <u>Moncure Robinson (1802-
1891)</u> (New York: Newcomen Society, American Branch,
1949), p. 18.
    42. <u>Reports and Documents Relative to the Termina-
tion of the Pennsylvania Rail Road</u> (Philadelphia: [n. p.],
1830), pp. 15-19; <u>Reports of Moncure Robinson, Esq. &
Col. Stephen H. Long, Engineers Appointed by the Canal
Commissioners for Examining the Different Routes for
Crossing the Allegheny Mountain</u> (Harrisburg, Pa.: 1831),
pp. 3-18.
    43. <u>Report of Moncure Robinson . . . Jonathan
Knight . . . Benjamin Wright . . . upon the Plan of the
New York and Erie Rail Road</u> (New York: George P. Scott
& Co., 1835). In <u>The American Civil Engineer</u>, pp. 78-
82, Daniel Calhoun discusses the "tasks of the engineer"
at length.
    44. Harry D. Jones to Moncure Robinson, 27 April
1837, MRP; The Directors of the Kaiser Ferdinand Nord-
bahn to Moncure Robinson, 1 May 1838, MRP; Baron de
Roenne to Moncure Robinson, 18 May 1840, MRP; Michel
Chevalier to the Prince of Austria, [c. 1843], MRP; [Rich-
ard Boyse Osborne], <u>Sketch of the Professional Biography of Moncure Robinson</u> (Philadelphia: J. B. Lippincott
Company, 1888), pp. 37-38.
    45. Moncure Robinson to John White, 5 April 1829,
Adelman Collection, HSP; Jno. Blair to Moncure Robinson,
28 March 1830, MRP; William Pope to Conway Robinson,
16 February 1836, MRP. The following quote is revealing:
". . . we must keep, Wirt, the game in Virginia in our
hands, & allow no one good, bad or indifferent to intrude.
Managing matters in our own way we can insure the suc-

cess of any project we may take charge of & improve this
nascent spirit which is beginning to develop itself in so
many directions in Va." (Moncure Robinson to Wirt Robinson, 2 October 1830, RC.)

46. E. g. , Moncure Robinson, Report of the Engineer,
on a Survey of a Railroad from Petersburg to the Roanoke
(Petersburg, : [n. p.], 1830), pp. 10-11; on engineers and
promotion, see Calhoun, The American Civil Engineer,
pp. 65-68.

47. Claudius Crozet did the original survey for the
Chesterfield Railroad: Claudius Crozet, "Report on the
Rail-Way from the Coal-Pits to James River," in Twelfth
Annual Report of . . . the Board of Public Works (Rich-
mond: Samuel Shepherd & Co. , 1828), pp. 339-47.

48. Moncure Robinson to Conway Robinson, 4 November 1838, MRP.

49. His relationship with at least two other railroads remains obscure. Robinson was Chief Engineer of
the Winchester (Va.) and Potomac Railroad in 1833-1834
while it was under construction, but it is not clear whether
he held that position until the road's completion. In 1836
Robinson was Consulting Engineer on the Gaston & Raleigh
Railroad in North Carolina, and in 1837 one of his former
assistants on the Philadelphia and Reading was in charge of
construction on the G & R. See: Moncure Robinson to Conway Robinson, 21 March 1833, MRP; Minutes of President
and Directors of the Winchester and Potomac Railroad Company, 28 June 1834, MRP; Wm. H. Morell to Moncure
Robinson, 7 July 1834, MRP; entries for 3 May 1836 and
9 December 1837, W. M. C. Fairfax diary, VHS; T. P.
Devereaux to Moncure Robinson, 26 February 1836; Michel
Chevalier, Histoire et Description des Voies de Communication aux États-Unis (Paris: Librairie de Charles Gosselin,
1840-1841), II:37; Niles' Weekly Register, 21 September
1833.

50. Francis Earle Lutz, Chesterfield: An Old Virginia County (Richmond: William Byrd Press, Inc. , 1954),
pp. 181-82; Wirt Robinson to James Brown, Jr. , 14 No-

vember 1830, box 15, VaBPW; Chevalier, Histoire et
Description des Voies de Communication aux États-Unis,
II:448, 522-24; Richmond Enquirer, 21 June 1831; Register of Pennsylvania VIII (1831):361; Moncure Robinson
to John Robinson, 29 June 1830 and 28 June 1831, MRP.

51. "Petersburg Railroad Company," in Eighteenth
Annual Report of . . . the Board of Public Works (Richmond: Samuel Shepherd & Co., 1835), pp. 182-83; Moncure Robinson to Wirt Robinson, 2 October [1830], RC;
Moncure Robinson to James Brown, Jr., 6 December 1832,
in Seventeenth Annual Report of . . . the Board of Public
Works (Richmond: Samuel Shepherd & Co., 1833), pp. 43-
47; Chevalier, Histoire et Description des Voies de Communication aux États-Unis, I:422.

52. Miners' Journal (Pottsville, Pa.), 28 May 1831;
Reports of the Engineers of the Danville & Pottsville Rail
Road Company . . . October 15, 1831 (Philadelphia: Clark
& Raser, 1831), pp. 9-13; Chevalier, Histoire et Descrip-
tion des Voies de Communication aux États-Unis, II:506-22;
John N. Hoffman, Girard Estate Coal Lands in Pennsylvania,
1801-1884 (Washington, D. C.: Smithsonian Institution Press,
1972), pp. 62-67, 69-76.

53. John V. Hare, History of the Reading (Philadelphia: John Henry Strock, 1966 [reprint]), pp. 2-4; [Osborne],
Sketch of the Professional Biography of Moncure Robinson,
p. 26.

54. Ibid., p. 23.

55. John B. Mordecai, A Brief History of the Richmond, Fredericksburg and Potomac Railroad (Richmond:
Old Dominion Press, Inc., 1941), pp. 8-9.

56. Second Meeting of the [Richmond and Petersburg]
Stockholders (Richmond: Thomas W. White, 1837), p. 33;
Proceedings of the Stockholders in the Richmond and Petersburg Rail Road Company . . . (Richmond: Thomas W. White,
1839), p. 61: Chevalier, Histoire et Description des Voies
de Communication aux États-Unis, II:417, 571-75.

57. This is apparent from the Wilson Miles Cary
Fairfax diary, VHS.

58. Cf. "List of Stockholders in the Richmond &

Petersburg Railroad Company . . . 16 Feby. 1837," box
46, VaBPW; Moncure Robinson to Conway Robinson, 21
February 1836, MRP.
59. Mordecai, Brief History of the Richmond, Fredericksburg and Potomac Railroad, p. 20.
60. "John Bloomfield Jervis," DAB V, 59-60; "John
Edgar Thomson," DAB IX, 486-87.
61. Moncure Robinson to Joseph McIlvaine, 30 July
1828, box 1, Surveys & Corr., PaCC.
62. Moncure Robinson to Wirt Robinson, 2 October
[1830], RC; Moncure Robinson to John Robinson, 4 November 1830, MRP; Calhoun, The American Civil Engineer, p. 212; Cullum, Biographical Register . . ., I :238-
39. 63. Moncure Robinson to Bd. of Pub. Wks. of Va.,
31 January 1835, in (uncatalogued) Moncure Robinson box,
VaBPW; Charles W. Turner, "The Louisa Railroad, 1836-
1850," North Carolina Historical Review XXIV (1947):37,
39; Chevalier, Histoire et Description des Voies de Communication aux États-Unis, II: 412.
64. "Richard Boyse Osborne, C.E.," Engineering
News XLII (1899):394. I thank Hugh Gibb of the Eleutherian
Mills Historical Library for this reference.
65. Wilson, Reminiscences of a Railroad Engineer,
pp. 44-55.
66. Eugene S. Ferguson, "Toward a Discipline of
the History of Technology," Technology and Culture XV
(1974):20-21.

COMMENTARY

John B. Rae

My first task as commentator is the pleasant one of congratulating our speakers on thoughtful and provocative discussions of two individuals whose careers are illuminating for the theme of this conference--and whether by accident or design, illuminate it from radically different perspectives. Latrobe was English, from a family deeply rooted in European culture, and he had the best professional training available in eighteenth century Britain. Robinson, the son of a Virginia merchant, had three years of college but no degree, and he learned his engineering on his own. Latrobe brought the latest in European technology with him in person when he came to the United States; Robinson first made himself an engineer and then went to Europe to learn more.

Both speakers have begun by formulating the premises on which they base their interpretations of how their subjects contributed to the transfer of technology. These premises focus on the personal factor, and I agree. What Mr. Carter terms the "mysteries" of the engineer's calling are best transmitted by direct communication, and in the period we are considering were almost exclusively transmitted in this way. I am not sure I would exclude indirect communication quite to the extent that Mr. Stapleton does. I lean to Eric Robinson's position that "the diffusion of technical knowledge is seldom confined to one route. All is grist to the mill of the enquiring mind."[1] But while there were alternative routes, there was never any question about which was the main highway.

If we apply this premise to the nineteenth century, Latrobe and Robinson personify two different methods of effecting the transfer of European technology to America. Latrobe represents the large number of European trained engineers who brought their knowledge with them, applied it, adapted it, and taught it. Some raw figures on this movement may be of value. I have just been through the

Biographical Archive of American Engineers at The
National Museum of History and Technology of the Smithsonian Institution and found 1,672 engineers about whom
reasonably adequate information was available. These
are American engineers born before 1860. Of these, 219,
or about 13 percent, were born in Europe, but allowance
has to be made for a large number brought over by their
parents while they were children. Education may be
more significant. Of the 906 who had formal engineering
education, 126, 14 percent, were educated abroad. This
includes both foreign-born engineers who completed their
training before coming to the United States and Americans
who sought engineering education in Europe. The Archive
is incomplete; the information is fragmentary; there are
unquestionably errors in my recording and my arithmetic,
but I believe these numbers are a close enough approxima-
tion to give a reasonably accurate picture of the general
situation.

We need to add also the American engineers like
Moncure Robinson and the others mentioned in these papers who went to Europe to observe for themselves.   I
am unable to give any estimate of numbers; to do it would
require more biographical data than has so far been com-
piled about American engineers, even in the Smithsonian
Archive.

For the region in which our host Foundation is particularly interested, several examples of the Latrobe-
type transfer exist, beginning, as Mr. Stapleton has
pointed out, with E. I. du Pont, who may have been a
chemist rather than an engineer but who most definitely
transferred a superior technology from Europe. Norman
Wilkinson has offered evidence that the suggestion of
establishing a powder works originated with Louis de
Tousard, a French army engineer who came to this country during the American Revolution and whom Dr. Wilkinson also regards as the real founder of the United States
Military Academy.[2]   Other Europeans would include
William Weston, an English engineer who was probably

trained by James Brindley and who was associated with
most of the important early canal projects in the eastern
United States; two English papermaking experts--Thomas
Oakes, who rebuilt the Gilpin paper mill in Wilmington in
1804, and Lawrence Greatrake, brought over as superin-
tendent of the Gilpin mill in order to employ the latest
British techniques, including the first use of chlorine instead of muriatic acid for bleaching; and Emile Geyelin,
a French engineer who settled in Philadelphia in 1848
and introduced the Jonval water turbine to the industries
of the Brandywine, with the active encouragement of
Alfred V. du Pont.[3]

The Robinson pattern of American engineers going
to Europe to observe for themselves and sometimes, as
he did, to make up for a lack of formal training was more
common in the period spanned by Latrobe's lifetime and
his own (from Latrobe's birth in 1764 to Robinson's re-
tirement in 1847). Among engineers in this category who
worked in this region are names as distinguished as that
of Moncure Robinson himself: William Strickland, perhaps Latrobe's greatest pupil; Loammi Baldwin, Jr.,
for a brief and stormy period chief engineer of the Union
Canal in Pennsylvania; Canvass White, Baldwin's suc-
cessor; and the several railroad engineers mentioned
by Mr. Stapleton. Some, like Baldwin and Robinson,
went on their own. Others were sent by the organizations they worked for in order to investigate specific
techniques for application in the United States, as Strickland and his pupil Samuel Kneass did for the Pennsylvania
Society for the Promotion of Internal Improvement and
Whistler, McNeill, and Knight for the Baltimore and
Ohio Railroad.

However, mere proliferation of names is a fruit-
less exercise. Our problem is to determine whether,
with Benjamin Henry Latrobe and Moncure Robinson as
models, the engineer's role as an agent in the transfer
of technology can be identified, and if so, what is transferred and how.

It is seldom easy to pinpoint the specific technique
that is transferred or communicated; the acquisition of
British papermaking techniques and technicians by the
Gilpins is a rather rare example. What is more likely
to be transferred is, to repeat Mr. Carter's felicitous
phrase, "the mysteries of his calling"--or we might use
the familiar American expression "know-how." This is
a process, as Mr. Stapleton has said, that really requires communication between persons who are in each
other's presence. We have plenty of examples in our
own time of a phenomenon that applied equally well in
the Latrobe-Robinson era; transplanting technology is
not successfully accomplished just by expecting the re-
cipient to follow the simple instructions on the package.
     Next we have to consider the place of the engineer
in this process of transfer. Could his function as the
agent of transfer be performed just as well by someone
else? I believe not, and to support this belief I wish to
submit two more general premises, both of which I consider to be valid:
     1. While the transfer of technology is best achieved
by interpersonal communication, an institutional structure is needed in order for the process to be effective in
producing applicable results. In other words, hit-or-
miss contacts among individuals, no matter how highly
trained and skilled, are unlikely to be productive in the
actual transfer and use of technology. In the examples
before us Latrobe was the product of an emerging orga-
nized professional body of British civil engineers, for
which his teacher, John Smeaton, was largely responsible. (I use the term "civil engineer" in its original
connotation as including all non-military engineering.
Latrobe's work in mechanical engineering has been described; this was characteristic of practically all early
engineers.) Robinson came from a less recognized and
recognizable group, but American engineers were becoming conscious of their identity--even if Loammi Baldwin, Jr. did write as late as 1826, "From all I see and

learn there are yet no civil engineers in the United States, "
presumably excepting himself. [5]

More importantly, Latrobe and Robinson were able
to apply their knowledge in the United States because
there was a developing structure of public and private
agencies concerned with projects for economic development which both increasingly required the skills of the
trained engineer and also provided the necessary outlet
for these skills. Nathan Rosenberg has emphasized the
role of specialized capital goods firms in both Great
Britain and the United States as instruments for institu-
tionalizing technological change, firms such as were
developed in this part of the country by William Sellers
and Matthias Baldwin. [6] He makes a persuasive case,
but for the first half of the nineteenth century in the
United States I would give at least as much weight to
the organizations that promoted canals, railroads, har-
bors, water supply and sanitation systems--in short,
W. W. Rostow's "social overhead capital. "

2. Technological change, or advance, or progress,
whichever term we may wish to use, occurs predomi-
nantly in the form of small increments contributed by a
great many talented individuals rather than through the
few spectacular inventions or breakthroughs by those we
recognize as geniuses. Here I call Dr. Rosenberg to
my assistance:

> Schumpeter accustomed economists to
> thinking of technical change as involving major
> breaks, giant discontinuities or disruptions
> with the past . . . . But technological change
> is also (and perhaps even more importantly)
> a continuous stream of unnumerable minor
> adjustments, modifications, and adaptations
> by skilled personnel. [7]

This "skilled personnel" consists for the most part,
although not exclusively, of engineers, because engineers

are the persons most concerned with planning and design-
ing our technologies and are therefore in the most stra-
tegic position to see and use opportunities for "adjust-
ments, modifications, and adaptations."

Thus it is not particularly important that neither
paper offers much in the way of novel techniques or
pieces of hardware that either Latrobe or Robinson could
be said to have transferred from Europe to America.
Both came to substantially the same conclusion: namely,
that what Latrobe and Robinson contributed to American
technology was fundamentally knowledge of superior techniques and higher levels of skill and precision. That they
did make such a contribution is unquestionable, but it is
not as clear how well they did in the matter of adaptability
to the differences between the British and the American
situations--and in our own time we have learned by hard
experience that adaptation is essential to the success of
any attempted transfer of technology.

On the record Robinson seems to come out somewhat
better than Latrobe, no doubt because he was American
and he received most of his education in this country. Mr.
Stapleton rates the building of the Philadelphia and Reading Railroad to contemporary British standards along a
difficult route as Robinson's greatest achievement, but
he does not suggest that Robinson tried to impose such
rigorous standards on other railroads whose prospects
for traffic were more chancy than the Reading's projected
75, 000 tons of anthracite a day.

Latrobe came in for a good deal of criticism. There
may be some local interest in the fact that he was attacked
in 1803 by Oliver Evans for a report stating that steam
vessels and steam wagons would need their entire carrying capacity just for engines and fuel--an appraisal not
absurdly out of line in 1803.[8] When Loammi Baldwin,
Jr. was on his way to his appointment on the Union Canal
in 1821, he noted in his journal:

> The old scheme for a canal from Dela-
> ware Bay to Chesapeake Bay is reviving.

> Everybody says it is easy and ought to be
> done. It was a fiasco under Latrobe many
> years ago. I see signs of his work and
> realize his folly. [9]

No explanation of the folly is offered, and coming as it
did just a year after Latrobe's death, the remark probably illustrates nothing more than Baldwin's customary
derogatory attitude toward his fellow engineers.

Mr. Carter concedes that Latrobe did at times have
trouble with the differences between British and native
American technology and was occasionally frustrated by
having to "compromise his fundamental aims." The con-
flict was between the British tradition of design for permanence and low upkeep and the American desire for low
investment and short amortization. In particular Latrobe
disliked the American propensity to use wood freely for
major structural purposes, a view shared by his pupil
William Strickland. Strickland made the cogent argu-
ment that the use of wood for structures such as canal
locks should be avoided if stone was convenient, because the need for frequent repairs was a serious impedi
ment to traffic. [10] On the other hand, David Stevenson,
a British engineer and Robert Louis Stevenson's uncle,
was much impressed by the way American engineers employed wood where Europeans would never have considered
anything but masonry and commended the Americans in the
highest terms for their ingenious adaptation of their techniques to the economic limitations of an undeveloped country
and the availability of materials. He is well worth quoting:

> At the first view, one is struck with the tem-
> porary and apparently unfinished state of many
> of the American works, and is very apt, be-
> fore inquiring into the subject, to impute to
> want of ability what turns out, on investigation,
> to be a judicious and ingenious arrangement to

suit the circumstances of a new country, of
which the climate is severe, -- a country
where stone is scarce and wood is plentiful,
and where manual labour is very expensive.
It is vain to look to the American works for
the finish that characterises those of France,
or the stability for which those of Britain are
famed. Undressed slopes of cuttings and
embankments, roughly built rubble arches,
stone parapet-walls coped with timber, and
canal-locks wholly constructed of that material, every where offend the eye accus-
tomed to view European workmanship. But
it must not be supposed that this arises from
want of knowledge of the principles of engineering, or of skill to do them justice in the
execution. The use of wood, for example,
which may be considered by many as wholly
inapplicable to the construction of canal-locks,
where it must not only encounter the tear and
wear occasioned by the lockage of vessels, but
must be subject to the destructive consequen-
ces of alternate immersion in water and ex-
posure to the atmosphere, is yet the result of
deliberate judgment. The Americans have, in
many cases, been induced to use the material
of the country, ill adapted though it be in some
respects to the purposes to which it is applied,
in order to meet the wants of a rising community, by speedily and perhaps superficially
completing a work of importance, which would
otherwise be delayed, from a want of the means
to execute it in a more substantial manner; and
although the works are wanting in finish,   and
even in solidity, they do not fail for many years
to serve the purposes for which they were constructed, as efficiently as works of a more last-
ing description. [11]

Mr. Carter has also seen Latrobe as an example of the engineer transferring his unique skills and knowledge from a sophisticated technology to a less advanced one. This was then and is now the customary pattern of technological transfer, but there is no "second law" in this field to say that the flow is irreversible, and in those days at least it was not.   To give just one example, the refugee French engineer Marc I. Brunel, later Sir Marc Brunel, spent five years working on engineering projects in the State of New York.   Then he left for England in 1799, carrying with him what he called "a cargo of ideas, " including the use of the circular saw on its timber carriage-- a Middle-European technology which had made its way to America but not to Britain and which in fact met resistance when Brunel first introduced it in the Portsmouth Navy Yard. [12] The principal part of the "cargo of ideas" was Brunel's project for the machine manufacture of ships' pulley blocks, which he succeeded in carrying out with the cooperation of Sir Samuel Bentham and Henry Maudslay, but this was a novel concept of Brunel's and not a technology already in existence. [13]

Another variation, which has its modern counterparts, is illustrated in the life of George Washington Whistler.   Mr. Stapleton has mentioned him as one of the three young engineers sent to Britain to study railway techniques for the Baltimore and Ohio.   Having started his career by helping to introduce British railway technology into the United States, he ended it by introducing American railway technology to Russia.   He began the first Russian railway line in 1842, between St. Petersburg and Moscow, and it was finished by his son after the senior Whistler died of cholera in 1849.   The son, George William Whistler, continued to build railroads in Russia for another twenty years.   The Whistlers illustrate a transfer from a less advanced technology to a still less advanced one, and it can be argued that American railway technology of the mid-nineteenth century was more applicable to contemporary Russian conditions than the more sophisticated

British technology would have been. The analogy with
present-day programs of technical assistance is, I hope,
self-evident.

       To sum up, what we have before us is not just an
intellectual exercise in historical research.   The careers of Benjamin Henry Latrobe and Moncure Robinson
illustrate, in an earlier and simpler setting, many of the
problems encountered in our own efforts to extend technical assistance to the less developed areas of the world.
The experience of two individuals is far from telling the
whole story.   There are other nineteenth century engineers, a number of them in this region, whose careers
would provide further opportunity to learn some lessons
from history that have a real bearing on issues that con-
front us now.   The alternative has been conclusively
stated by George Santayana: "Those who remain ignorant
of history are condemned to repeat it."

FOOTNOTES

1. Eric Robinson, "The Early Diffusion of Steam Power," Journal of Economic History 34 (March 1974):94.

2. Norman B. Wilkinson, "The Forgotten Founder of West Point," Military Affairs 24 (Winter 1960-61):184.

3. - 4. H. B. Hancock and N. B. Wilkinson, "Thomas and Joshua Gilpin: Papermakers," The Paper Maker 29 (1958):5; Roy M. Boatman, Turbines, Hagley Museum Research Report, 1956, p. 26.

5. F. K. Abbott, The Role of the Civil Engineer in Internal Improvements: The Contributions of the Two Loammi Baldwins, Father and Son, 1776-1838 (Ann Arbor, Mich.: University Microfilms, 1952). (Xerox copy in Eleutherian Mills Historical Library.)

6. Nathan Rosenberg, "Economic Development and the Transfer of Technology," Technology and Culture 11 (October 1970):557.

7. Ibid., 568-69.

8. Eugene S. Ferguson, ed., Early Engineering Reminiscences of George Escol Sellers, U. S. National Museum Bulletin 288 (Washington, D. C.: Smithsonian Institution, 1965), p. 37.

9. Abbott, Civil Engineer, 110.

10. William Strickland, Reports on Canals, Railways, Roads and Other Subjects to the Pennsylvania Society for the Promotion of Internal Improvement, (Philadelphia; H. C. Carey & I. Lea, 1826), 11.

11. David Stevenson, The Civil Engineering of North America (London: J. Weale, 1838), 192-93.

12. Paul Clements, Marc I. Brunel (London: Longmans, 1970), 47.

13. Another and perhaps stronger motive for going to England was Sophie Kingdom, whom Brunel had met in France while both were fugitives from the Terror and who became the mother of Isambard Kingdom Brunel.